An Anatomical Wordbook

To Annette

An Anatomical Wordbook

Stephen Lewis DCR(R) BSc PhD

Prosector, Department of Anatomy
University of Wales College of Cardiff

Butterworth–Heinemann

London   Boston   Toronto   Singapore   Sydney   Wellington

PART OF REED INTERNATIONAL P.L.C.

First published 1990

© Butterworth–Heinemann Ltd, 1990

British Library Cataloguing in Publication Data
Lewis, Stephen
    An anatomical wordbook.
    1. Man. Anatomy
    I. Title
    611
    ISBN 0-7506-1023-9

Library of Congress Cataloging in Publication Data
Lewis, Stephen, Ph.D.
    An anatomical wordbook/Stephen Lewis.
        p.   cm.
    ISBN 0-7506-1023-9:
    1. Human anatomy – Terminology.   I. Title.
    QM7.L49   1990
    611'.0014 – dc20   90-2540 CIP

Text processed by David Coates, Tunbridge Wells, Kent
Composition by Genesis Typesetting, Laser Quay, Rochester, Kent
Printed and bound by Hartnolls Ltd., Bodmin, Cornwall

Preface

Anatomy is a science which, to many of its students, is couched in words which are difficult to understand, being drawn from other languages, cultures and eras. Many have found a breakthrough in comprehension of the subject through a fuller understanding of its terminology since, very often, the words used are merely simple descriptions of anatomical structures but in an unfamiliar language. If, as it is said, 'every picture tells a story' then happily in Anatomy 'every word is a picture'.

It is the aim of this book, therefore, to provide all who study Anatomy with a simple reference guide to such words in order that their comprehension of the subject may be enhanced.

This is not simply a dictionary of anatomical words, although there is a central alphabetical Glossary. Rather, numerous words have been omitted from this Glossary and grouped together by common themes into a series of relatively short sections at the front of the book. In this way, cross-referencing within these themes is made much easier since all related entries are in one place. The list of contents is, therefore, a very necessary indicator and guide to the location of a word. Numerous anatomical structures additionally bear the name of a person with whom they have become associated. Although the use of eponyms is now discouraged, their use is not extinct and so a list is provided at the end.

The boundaries of Anatomy are not clearly defined and there is often overlap, especially within the spheres of cell and molecular biology. For the purposes of this book, however, words pertaining to intra- and sub-cellular features have, by and large, been avoided. This is appropriate given that the vast majority of those words in need of explanation significantly pre-date the use of microscopy, relating instead to gross anatomy.

This book started out as a collection of explanations to anatomical words kept for personal interest. When it was seen how helpful knowledge of these explanations could be to others, the collection was developed with a view to wider use. This project would not have reached as satisfactory a conclusion, however, had it not been for the assistance of my wife, Annette, whose influence is implicit throughout and to whom this work is dedicated.

Contents

Preface   v
Abbreviations   ix
Common prefixes   1
Common suffixes   5
Some anatomical words ending in -oid   7
Some anatomical words ending in -iform   9
Some anatomical words ending in -ate   10
Words pertaining to number and quantity   11
Anatomical words where Latin and Greek synonyms are often used   12
Names applied to muscles   14
Bone names and their meanings   22
Names of the cranial nerves and their meanings   26
Words pertaining to joint types   28
Words pertaining to action, movement and position in space   31
Some anatomical lines, planes and points   34
The regions of the abdomen   37
Named points on the skull   38
Head types   40
Body types   41
Glossary   A   42
         B   49
         C   51
         D   62
         E   65
         F   70
         G   73
         H   75
         I   77
         J   80
         K   80
         L   80
         M   83
         N   89
         O   91
         P   94

Q	**103**
R	**103**
S	**105**
T	**111**
U	**116**
V	**118**
Z	**121**

Eponyms   122

Abbreviations

(Ar)	Arabic (5th–15th Cent.)
(Dut)	Dutch (from 16th Cent.)
(Fr)	French (from 17th Cent.)
(Gk)	Greek (Classical) (5th–1st Cent. BC)
(Icel)	Icelandic (from 13th Cent.)
(L)	Latin (Classical) (1st Cent. BC–5th Cent. AD)
(LL)	Late Latin (5th–15th Cent.)
(ME)	Middle English (12th–15th Cent.)
(Med Gk)	Medieval Greek (4th–15th Cent.)
(Mod L)	Modern Latin (from 16th Cent.)
(OE)	Old English (5th–12th Cent.)
(OF)	Old French (12th–15th Cent.)

(In accordance with contemporary thinking, the term 'Old English' is used synonymously with and in place of 'Anglo-Saxon'.)

Common prefixes

a-, an-	(Gk): absence of, without, not (Alpha privatum)
ab-	(L): away from
ad-	(L): to
ambi-	(L): round about, both
amphi-	(Gk): on both sides
ana-	(Gk): up, through
andro-	(Gk): male
aniso-	(Gk): unequal
ante-	(L): before
anti-	(Gk): opposite
apo-	(Gk): from
arche-, archi-	(Gk): first
arthro-	(Gk): joint
auto-	(Gk): self
bi-	(L): twice
bio-	(Gk): life
brachy-	(Gk): short
cata-	(Gk): motion down from
chondro-	(Gk): cartilage
circum-	(L): around
co-, con-	(L): together, with
contra-	(L): against
crypto-	(Gk): hidden
cum-	(L): with
cyst-	(Gk): bladder
cyto-	(Gk): cell
de-	(L): down, away, from
demi-	(L): half, lesser
desmo-	(Gk): membrane
deutero-	(Gk): second
di-, dis-	(Gk): twice, double
di-, bis-	(L): in two, apart
dia-	(Gk): through
dolicho-	(Gk): long

1

dys-	(Gk): bad, defective, difficult, painful (English mis-)
e-, ec-	(Gk): out from
ecto-	(Gk): outside
em-, en-	(Gk): in, within, on
endo-	(Gk): within
ento-	(Gk): within
epi-	(Gk): upon
eu-	(Gk): well
ex-	(L): out
exo-	(Gk): outside
extra-	(L): outside, in addition to, more
haema-	(Gk): blood
hemi-	(Gk): half
hetero-	(Gk): different, the other of two
histo-	(Gk): tissue
holo-	(Gk): whole
homo-, homeo-	(Gk): same, like
hyper-	(Gk): over, in
hypo-	(Gk): below, deficient
hystero-	(Gk): uterus
im-, in-	(L): in, not
infra-	(L): beneath
inter-	(L): between
intra-	(L): within
is-, iso-	(Gk): equal, the same
juxta-	(L): adjoining
karyo-	(Gk): pertaining to a nucleus
kerato-	(Gk): horny
leuco-	(Gk): white
macro-	(Gk): long, large
mast-	(Gk): breast
mega-	(Gk): big
melan-	(Gk): black
men-	(Gk): month
meso-	(Gk): middle, among
meta-	(Gk): after, among, beyond, between
metra-	(Gk): uterus
micro-	(Gk): small

mono-	(Gk): one, alone, sole
myel-	(Gk): marrow, brain or spinal medulla
myo-	(Gk): muscle
neo-	(Gk): new
nephr-	(Gk): kidney
neuro-	(Gk): nerve (or sinew)
omo-	(Gk): shoulder
ophthalm-	(Gk): eye
ortho-	(Gk): straight
osteo-	(Gk): bone
oto-	(Gk): ear
oxy-	(Gk): sharp
pan-	(Gk): all
paleo-	(Gk): old
para-	(Gk): beside
per-	(L): through
peri-	(Gk): around
platy-	(Gk): wide
pleo-	(Gk): more
poikilo-	(Gk): unequal
poly-	(Gk): many
post-	(L): after, behind
pre-, prae-	(L): prior, before, in front of
pro-	(Gk): prior, before, in front of
procto-	(Gk): anus
proto-	(Gk): first
pseudo-	(Gk): false
pylo-	(Gk): pelvis
re-	(L): again, turning back
retro-	(L): behind
rubro-	(L): red
sarco-	(Gk): flesh
semi-	(L): half
sub-	(L): under
super-	(L): over, above
supra-	(L): over, above (a place)
sym-	(Gk): with, together
syn-	(Gk): together with
ter-	(L): three

trans-	(L): across
tri-	(Gk): three
ultimo-	(L): ultimate, furthest, last
uni-	(L): one
vaso-	(L): vessel

Common suffixes

-aemia	(Gk): pertaining to blood
-agogue	(Gk): to lead or carry off
-agra	(Gk): a seizure
-algia	(Gk): pain
-blast	(Gk): pertaining to formative cells
-coele, -cele	(Gk): cavity, rupture
-cyte	(Gk): pertaining to (adult) cells
-ectomy	(Gk): a cutting out
-eous	(Gk): of that kind
-fer	(L): a carrier of
-graph	(Gk): to write, describe
-ia	(Gk): the morbid condition of the root word
-ic	(Gk): used to form an adjective and so meaning to pertain to the root word
-iculus	(L): the diminutive of the root word (Anglicized to -icle)
-igo	(L): to do or to act
-ismus	(Gk): used to denote disease
-itis	(Gk): has come to mean inflammation
-logy	(Gk): treatise, word
-mania	(Gk): madness
-odynia, -odyne	(Gk): pain
-oid	(Gk): form
-oma	(Gk): pertaining to a tumour
-opia	(Gk): eye or sight
-osis	(Gk): a condition pertaining to the root word
-ostomy	(Gk): entrance, opening
-ous	(L): of that kind
-pathy	(Gk): suffering
-phobia	(Gk): fear
-plasty	(Gk): to form or mould
-rhagia	(Gk): to burst forth
-rhoea	(Gk): pertaining to a flow

-sis	(Gk): used to denote a process, action or progression
-scope, -scopy	(Gk): pertaining to viewing and making visible
-smus	(Gk): used to make verbal nouns
-tomy	(Gk): to cut
-ulus	(Gk): the diminutive of the root word
-uria	(Gk): pertaining to urine
-yl	(Gk): matter, stuff

Some anatomical words ending in -oid

The ending -oid (from eidos (Gk): shape or form) makes the resulting word mean like or of the form indicated by the root word e.g. cuboid means cube-like or cube-shaped.

Adenoid	Gland
Alisphenoid	Wing and wedge
Amygdaloid	Almond
Arachnoid	Spider's web
Arytenoid	Cup or ladle
Choroid	Leather or parchment
Clinoid	Bed (Four poster)
Colloid	Glue
Conoid	Cone
Coracoid	Crow (especially its beak)
Cotyloid	Cup or pan
Cricoid	(Signet) Ring
Cuboid	Cube
Deltoid	The Greek letter — Δ (Delta)
Desmoid	Band or ligaments
Discoid	Disc
Ethmoid	Sieve
Fibroid	Thread, fibre
Glenoid	Shallow socket
Hyaloid	Glass
Hyoid	The Greek letter — υ (Upsilon)
Lambdoid	The Greek letter — λ (Lambda)
Lipoid	Fat
Mastoid	Breast
Odontoid	Tooth (peg)
Osteoid	Bone
Ovoid	Oval
Pterygoid	Wing
Rhomboid	Rhombus (diamond-shaped)
Scaphoid	Boat

Sesamoid	Seed
Sigmoid	The Greek letter — ς (Sigma)
Sinusoid	Sinus
Sphenoid	Wedge
Styloid	Pillar or pen
Thyroid	Shield (A long shield with a notch at the top for the user's chin)
Trapezoid	Small table
Xiphoid	Sword

Compound words ending in -oid

Geniohyoid	The muscle running from the chin to the hyoid bone
Mylohyoid	The muscle running from the mandible to the hyoid bone
Omohyoid	The muscle running from the shoulder to the hyoid bone
Parathyroid	(Glands) lying beside the thyroid gland

Some anatomical words ending in -iform

The ending -iform (from forma (L): shape or form) makes the resulting word mean like or of the form indicated by the root word e.g. piriform means pear-like or pear-shaped.

Bipenniform	Twice feathered
Bulbiform	Bulb
Cordiform	Heart
Cribriform	Sieve
Cuneiform	Wedge
Cymiform	Boat
Emboliform	Wedge or plug
Ensiform	Sword
Falciform	Scythe
Filiform	Thread
Fundiform	Sling
Fungiform	Mushroom
Fusiform	Spindle
Globiform	Globe
Lentiform	Lens (or lentil)
Mammilliform	Breast
Moniliform	String of pearls
Mytiform	Mussel
Pampiniform	Tendril
Penniform	Feather (also pennate)
Piriform	Pear
Pisiform	Pea
Pyriform	Pear
Reniform	Kidney
Restiform	Rope
Unciform	Hook
Vermiform	Worm
Ypsiliform	Y- or U-shaped υ (Upsilon) shaped

Some anatomical words ending in -ate

The ending -ate makes the resulting word mean like or of the form indicated by the root word e.g. dentate means toothed.

Arcuate	Bowed
Bicornuate	Having two horns
Bipennate	Having two feather-shaped structures
Capitate	Having a head
Caudate	Having a tail/tailed
Circumvallate	Having a round wall
Conjugate	Paired or twinned
Cordate	Heart-shaped
Corniculate	Being shaped like a small horn
Crenate	Notched
Cruciate	Crossed
Decussate	Cross
Dentate	Toothed
Foliate	Leaf-shaped
Geniculate	Having a knee or kink
Globate	Globular/spherical
Hamate	Hooked
Innominate	Unnamed
Laciniate	Having a fringe or a flap
Lobate	Having a lobe
Lobulate	Having a small lobe
Lunate	Being moon-shaped
Pennate	Being feather-shaped
Stellate	Being star-shaped
Turbinate	Coiled or shaped like a whorl
Uncinate	Hooked
Unipennate	Having one feather-shaped structure
Vallate	Walled

Words pertaining to number and quantity

English	Latin	Greek
One	uni-	mono·
Two	bi-	di·
Three	ter-	tri·
Four	quadri-	tetra·
Five	quinque-	penta·
Six	sexa-	hexa·
Seven	septa-	hepta·
Eight	octo-	octo·
Nine	novem-	ennea·
Ten	decem-	deka·
Eleven	undecem-	endeka·
Twelve	duodecem-	dodeka·
First	prim-	proto·
Second	secund-	deuter·
Third	tert-	trit·
One hundred	centi-	hecto·
One thousand	milli-	kilo·
Half	semi-	hemi·
One half more	sesqui-	
Whole	omni-	holo·
Equal	equi-	homo·
Many	multi-	poly·
More	super- and per-	hyper
Less	sub-	hypo

Anatomical words where Latin and Greek synonyms are often used

English	Latin	Greek
Anus	Anus	Proctos
Body	Corpus	Soma
Bone	Os (Ossis)	Osteon
Blood	Sanguis	Haima
Brain	Cerebrum	Encephalon
Breast	Mamma	Mastos
Cartilage	Cartilago	Chondros
Cell	Cella	Cytos
Chest	Thorax	Stethos
Clavicle	Clavicula	Cleis, Cleidos
Ear	Auris	Ous (Otos)
Elbow	Cubitum	Ancone
Eye	Oculus	Ophthalmos
Eyelid	Palpebra	Blepharon
Finger or Toe	Digitus	Dactylos
Flesh	Caro, Carnis	Creas, Sarx
Foot	Pes (Pedis)	Pous (Podos)
Gland	Glans	Aden
Glans Penis	Glans Penis	Balanos
Gum	Gingiva	Oulon
Hand	Manus	Cheir
Head	Caput, -ceps	Cephale
Heart	Cor	Cardia
Hip	Coxa	Ischion
Intestine	Intestinus	Enteron
Jaw	Maxilla, Mandibulum	Gnathos
Joint	Articulatio	Arthron
Kidney	Ren	Nephros
Knee	Genu	Gonu
Leg	Crus	Cneme
Lip	Labrum	Cheilos
Lung	Pulmo	Pleumon, Pneumon

Marrow	Medulla	Myelos
Mouth	Os (oris), Bucca	Stoma
Muscle	Musculus	Mys
Nail	Unguis	Onyx
Navel	Umbilicus	Omphalos
Neck	Cervix	Trachelos
Nerve	Nervus	Neuron
Nose	Nasus	Rhis (Rhinos)
Omentum	Omentum	Epiploon
Organ	Viscus	Organon
Ovary	Ovarium	Oophoron
Ovum	Ovum	Oon
Pelvis	Pelvis	Pyelos
Penis	Penis	Phallos
Pubis	Pubis	Epision
Shoulder	Humerus	Omos
Skin	Cutis, Corium	Derma, Pella
Spine	Spina	Rhachis
Spleen	Lien	Splen
Tail	Cauda	Cercos
Tendon	Tendo	Tenon
Testicle	Testis	Orchis
Tibia	Tibia	Perone
Tooth	Dens (Dentis)	Odous (Odontos)
Tongue	Lingua	Glossa, Glotta
Uterus	Uterus	Hystera, Metra, Delphys
Vagina	Vagina	Colpos
Vein	Vena	Phleps
Vertebra	Vertebra	Spondylos
Viscera	Viscera	Splanchna

Names applied to muscles

Elements of a muscle's action, position, shape etc. are often implied by its name. In order to help facilitate translation of a muscle name into English the following list of words applied to muscles and their meanings is given. By chaining together the meanings of each word in the muscle's name and then reordering them where appropriate, a more readily remembered description of the muscle is obtained e.g. flexor digitorum superficialis renders Flexor/Of the fingers/Superficial. Upon reordering, the muscle name reads: 'The superficial flexor of the fingers'.

Abdominis	Of the abdomen
Abductor	Mover *away* (from reference plane)
Accessorius	Accessory
Adductor	Mover *towards* (reference plane)
Anal	Of the anus
Anconeus	The elbow M.
Anguli	Of the angle
Anterior	Anterior
Antitragicus	The M. of the antitragus
Arrector pili	An erector of a hair (plur. arrectores pilorum)
Articularis genu	The knee articulator
Aryepiglotticus	The arytenoid (cartilage)-to-epiglottis M.
Arytenoid	(of the) Arytenoids
Auriculae	Of the ear
Auricularis	Ear M.
Biceps	The two-headed M.
Brachialis	The arm M.
Brachii	Of the arm
Brachioradialis	The arm-to-radius M.
Brevis	Short
Bronchoesophageus	The bronchus-to-oesophagus M.
Buccinator	The trumpeter M.

14

Bulbocavernosus	The bulb-to-corpus cavernosum (of penis or clitoris) M.
Bulbospongiosus	The bulb-to-corpus cavernosum (of penis or clitoris) M.
Capitis	Of the head
Carpi	Of the wrist
Cervicis	Of the neck
Chondroglossus	The cartilage-to-tongue M. From lesser horn of the hyoid bone to the tongue. (Considered by some to be part of hyoglossus.)
Coccygeus	M. of the coccyx
Colli	Of the neck
Compressor	Compressor
Constrictor	Narrower
Coracobrachialis	The coracoid-to-arm M.
Corrugator	Wrinkler
Costarum	Of the ribs
Cremaster	The suspender (or hammock) M.
Crico-arytenoid	The cricoid (cartilage)-to-arytenoid (cartilage) M.
Cricopharyngeus	The cricoid (cartilage)-to-pharynx M.
Cricothyroid	The cricoid (cartilage)-to-thyroid (cartilage) M.
Dartos	'Flayed' or 'skinned'
Deep	Deep (further from the surface)
Depressor	Lowerer, flattener
Deltoid	The delta- (triangle-) shaped M.
Detrusor	The thruster out; the (urine) expeller M.
Diaphragm	'A dividing wall'
Digastric	The two bellied M.
Digiti	Of the finger/Of the toe
Digitorum	Of the fingers/Of the toes
Dilator	Enlarger
Dorsal	Of the dorsal aspect
Dorsi	Of the back
Epicranius	The M. upon the head

Erector spinae	The erector (straightener) of the spine
Extensor	Extensor, straightener
External	External, outside
Externus	External, outside
Fascia latae	The broad fascia (of the thigh)
Femoris	Of the thigh
Flexor	Flexor, bender
Gastrocnemius	The belly of the leg M.
Gemellus	A twin
Genioglossus	The chin-to-tongue M.
Geniohyoid	The chin-to-hyoid (bone) M.
Gluteus	The buttock
Gracilis	The slender M.
Hallucis	Of the great (big) toe
Helicis	Of the helix
Hyoglossus	The hyoid (bone)-to-tongue M.
Iliacus	The iliac M.
Iliocostalis	The ilium-to-rib M. (Lateral muscle group of erector spinae)
Iliocostocervicalis	The ilium-to-rib-to-neck M. (Lateral muscle group of erector spinae)
Incisivus	Incisive
Indicis	Of the index finger
Inferior	Lower, below
Infraspinatus	The M. below the spine (of the scapula)
Innermost	Innermost
Intercostal	Between the ribs
Intermedius	Intermediate
Internal	Internal, inside
Internus	Internal, inside
Interossei	Between bones
Interspinal	Between adjacent vertebral spinous processes
Intertransverse	Between vertebral transverse processes

Ischiocavernosus	The ischium-to-corpus cavernosum (of penis or clitoris) M.
Labii superioris	Of the upper lip
Labii superioris alaeque nasi	Of the upper lip and wing of the nostril
Lata	Wide
Lateral	Lateral
Lateralis	Lateral
Latissimus	Widest
Levator	Lifter
Levatores	Lifters
Longissimus	Longest. (Intermediate muscle group of erector spinae)
Longitudinal	Longitudinal
Longus	Long
Lumborum	Of the loin (or lumbar vertebrae)
Lumbrical	The worm (shaped) M.
Magnus	Great
Major	Greater
Manus	Of the hand
Masseter	The chewer M.
Maximus	Greatest
Medial	Medial
Medialis	Medial
Medius	Middle
Mentalis	The chin M.
Middle	Middle
Minimi	Small
Minimus	Smallest
Minor	Smaller
Multifidus	The much divided M.
Musculi	Muscles
Musculus uvulae	The muscle of the uvula
Mylohyoid	The mandible-to-hyoid (bone) M.
Naris	Of the nose
Nasalis	The nose M.
Oblique	Oblique
Obliquus	Oblique
Obturator	M. of the obturator membrane

Occipitofrontalis	The occipital-frontal M.
Oculi	Of the eye
Omohyoid	The shoulder-to-hyoid (bone) M.
Opponens	Opposing
Orbicularis	Circular
Orbitalis	Of the orbit
Oris	Of the mouth
Palatoglossus	The palate-to-tongue M.
Palatopharyngeus	The palate-to-pharynx M.
Palmar	Of the palm of the hand
Palmaris	Of the palm of the hand
Palpebrae	Of the eyelid
Papillary	Nipple-like
Pectinati	Comb-like
Pectineus	M. of the pecten of the pubis
Pectoralis	Of the breast (or chest)
Pedis	Of the foot
Perinei	Of the perineum
Peroneus	Fibula
Piriformis	The pear-shaped M.
Plantar	Of the sole (plantar aspect) of the foot
Plantaris	The sole-of-the-foot M. (NB apparently named because it plantar flexes the foot.)
Platysma	The plate-like M.
Pleuroesophageus	The pleura-to-oesophagus M.
Pollicis	Of the thumb
Popliteus	The knee M. (N.B. popliteus properly refers to the Ham.)
Posterior	Posterior
Procerus	Extended or long
Profundus	Deep (further from the surface)
Pronator	Pronator (inward rotator of the forearm)
Prostatae	Of the prostate
Psoas	The loin M.
Pterygoid	Of the pterygoid process
Pubococcygeus	The pubis-to-coccyx M.

Puboprostaticus	The pubis-to-prostate (gland) M.
Puborectalis	The pubis-to-rectum M.
Pubovaginalis	The pubis-to-vagina M.
Pubovesicalis	The pubis-to-bladder M.
Pupillae	Of the pupils
Pyramidalis	The pyramid-shaped M.
Quadratus	Four-sided
Quadriceps	Four-headed
Radialis	Of the radius
Rectus	Straight
Rectococcygeus	The rectum-to-coccyx M.
Recto-urethralis	The rectum-to-urethra M.
Recto-uterinus	The rectum-to-uterus M.
Rectovesicalis	The rectum-to-bladder M.
Rhomboid	Rhomb-shaped (diamond-shaped)
Risorius	The 'laughter' M. (Properly, Risorius means 'laughable', so is a misnomer.)
Rotator	Rotator
Sacrococcygeus	The sacrum-to-coccyx M.
Sacrospinalis	The sacrum-to-spine M. (erector spinae)
Salpingopharyngeus	The (auditory) tube-to-pharynx M.
Sartorius	The tailor's (leg-crossing) M.
Scalenus	Uneven (of uneven side lengths)
Scapulae	Of the scapula
Semimembranosus	The half-membranous M.
Semispinalis	Partly between vertebral spinous processes (transverse-to-spinous processes infero-superiorly)
Semitendinosus	The half-tendinous M.
Septi	Of the (nasal) septum
Serratus	Serrated (saw-toothed)
Soleus	The sole (flatfish)-shaped M.
Sphincter	A binder or strangler
Spinalis	Between vertebral spinous processes (Medial muscle group of erector spinae)

Splenius	Bandage-like
Stapedius	The M. of the stapes
Sternocleidomastoid	The sternum/clavicle-to-mastoid process M.
Sternohyoid	The sternum-to-hyoid (bone) M.
Sternothyroid	The sternum-to-thyroid (cartilage) M.
Styloglossus	The styloid process-to-tongue M.
Stylohyoid	The styloid process-to-hyoid (bone) M.
Stylopharyngeus	The styloid process-to-pharynx M.
Subclavius	The M. below the clavicle
Subcostal	Beneath a rib
Subscapularis	The M. beneath the scapula
Supercilii	Of the eyebrow
Superficial	Superficial (nearer the surface)
Superficialis	Superficial (nearer the surface)
Superior	Higher, above
Superioris	Higher, above
Supinator	Supinator (outward rotator of the forearm)
Supraspinatus	The M. above the spine (of the scapula)
Temporalis	The M. of the temple
Temporoparietalis	The temporal-parietal M.
Tensor	Tightener
Teres	Round (NB none of the teres muscles are, in fact, round!)
Tertius	Third
Thoracic	Of the thorax
Thoracis	Of the thorax
Thyro-arytenoid	The thyroid (cartilage)-to-arytenoid (cartilage) M.
Thyro-epiglottic	The thyroid (cartilage)-to-epiglottis M.
Thyrohyoid	The thyroid (cartilage)-to-hyoid (bone) M.

Thyropharyngeus	The thyroid (cartilage)-to-pharynx M.
Tibialis	Of the tibia
Trachealis	The M. of the trachea
Tragicus	The M. of the tragus
Transverse	Transverse
Transversus	Transverse
Transversospinalis	(The obliquely running deep muscle group of the back)
Trapezius	The table-shaped M.
Triceps	Three-headed
Tympani	Of the tympanic membrane
Ulnaris	Of the ulna
Urethrae	Of the urethra
Vaginae	Of the vagina
Vastus	Vast, extensive
Veli palatini	Of the veil of the (soft) palate
Vertical	Vertical
Vesicae	Of the bladder
Vocalis	Of the voice
Zygomaticus	Zygoma M.

Bone names and their meanings

The words in this section are arranged regionally rather than alphabetically.

The Head and Neck

Occipital (bone)	The back of the head (bone). (From occipitium (L): the back of the head.)
Sphenoid (bone)	The wedge-shaped (bone). (From sphen (Gk): wedge + eidos (Gk): shape.)
Temporal (bone)	The (bone) of the temple. (From tempus (L): time.)
Malleus	Hammer. (From malleus (L): a hammer.)
Incus	Anvil. (From incus (L): an anvil.)
Stapes	Stirrup. (From stapes (L): a stirrup.)
Parietal (bone)	The (bone) of the walls (of the head). (From paries (L): a wall.)
Frontal (bone)	The (bone) of the front (of the head). (From frons (L): the forehead.)
Ethmoid (bone)	The sieve-like (bone). (From ethmos (Gk): sieve + eidos (Gk): shape.)
Inferior Nasal Conchae **or** **Inferior Turbinate(d) (bone)**	The lower shell(-like) (bone) of the nose. (From concha (L): a shell / kongche (Gk): a shell.) The lower coiled (bone). (From turbo (L): a whirl.)
Lacrimal (bone)	The (bone) of the tears. (From lacrima (L): a tear.)
Nasal (bone)	The (bone) of the nose. (From nasus (L): the nose.)
Vomer	A ploughshare (shaped bone). (From vomer (L): ploughshare.)
Maxilla	The (upper) jaw (bone). (From maxilla (L): jaw.)

Formerly called **Malar**	The cheek (bone). (From mala (L): the cheek.)
Palatine (bone)	The (bone) of the palate. (From palatum (L): the palate.)
Zygoma or Zygomatic (bone)	The (bone) connecting or yoking (bone) (between maxillary, frontal and temporal bones). (From zygoma (Gk): a connecting bar or yoke.)
Mandible	The chewing bone. (From mandere (L): to chew.)
Hyoid (bone)	The U- (υ) shaped (bone). (From hyoeides (Gk): U-shaped.)

The Spine and Chest

Vertebra	(L): A joint.
The cervical vertebrae	The neck vertebrae. (From cervix (L): neck.)
Atlas	Named after the Greek mythological character who supported the earth on his shoulders.
Axis	(L): a pivot.
The thoracic vertebrae	The chest vertebrae. (From thorax (Gk): a piece of armour for the chest and abdomen.)
The lumbar vertebrae	The loin vertebrae. (From lumbus (L): the loin.)
The sacrum	It is not clear why this bone is so named. (From sacer (L): sacred, holy.)
The coccyx	It is not clear why this bone is so named. (From kokkyx (Gk): cuckoo.)
The sternum	The breast bone. (From sternon (Gk): The breast of a man (not a woman).)
Manubrium	(L): A handle or hilt.
Body (gladiolus)	(L): (A small sword).
Xiphoid	Sword-shaped. (From xiphos (Gk): sword + eidos (Gk): shape.)
Ribs	From ribb (AS): rib

The Upper Limb

Scapula	(L): Shoulder-blade.
Humerus	From humerus (L): the shoulder.
Radius	A spoke of a wheel. (From radius (L): a wheel spoke.)
Ulna	The forearm. (From ulna (L): forearm, elbow.)
Carpus	The wrist. (From karpos (Gk): the wrist.)
Scaphoid	Boat-shaped. (From skaphe (Gk): a skiff, scapha (L): anything hollowed out + eidos (Gk): shape.)
Lunate	Moon-shaped (cresentic). (From luna (L): moon.)
Triquetral	Three-cornered. (From triquetrus (L): having three corners.)
Pisiform	Pea-shaped. (From pisum (L): pea + forma (L): shape.)
Trapezium	Four-sided. (From trapezion (Gk): a small table.)
Trapezoid	Four-sided, like a small table. (From trapezion (Gk): a small table + eidos (Gk): shape.)
Capitate	Head-shaped. (From caput (L): head.)
Hamate	Hooked. (From hamatus (L): hooked.)

Former names of some of the carpal bones

Scaphoid	Navicular
Lunate	Semilunar
Triquetral	Cuneiform
Pisiform	(None)
Trapezium	Greater Multangular
Trapezoid	Lesser Multangular
Capitate	Os Magnum
Hamate	Unciform
Metacarpals	After-the-wrist. (From meta (Gk): after + karpos (Gk): the wrist.)

Phalanx (phalanges)	From phalanx (Gk): 'a band of soldiers'.

The Lower Limb

The innominate bone	The unnamed bone.
Ilium	From ilium (L): the flank.
Ischium	From ischion (Gk): the socket for the femoral head.
Pubis	From pubes (L): the signs of manhood (i.e. pubic hair).
Femur	(L): Thigh.
Patella	(L): A small pan, plate or dish.
Tibia	(L): A tubular, musical wind instrument such as a flute.
Fibula	(L): A pin
Tarsus	(L): The ankle
Talus	(L): The ankle-bone.
Calcaneus	From calx (L): a heel.
Navicular	From navicula (L): a small boat.
Cuneiform	From cuneus (L) + forma (L): wedge-shaped.
Medial	
Intermediate	
Lateral	
Cuboid	From kuboeides (Gk): cube-shaped.

Former names of some of the tarsal bones

Talus	Astragalus (Gk): a die
Calcaneus	Os Calcis
Navicular	Tarsal Scaphoid
Cuneiform	Cuneiform
Medial	Internal
Intermediate	Middle
Lateral	External
Cuboid	(None)
Metatarsals	After-the-ankle. (From meta (Gk): after + tarsus (L): the ankle.)
Phalanx (phalanges)	See above.

Names of the cranial nerves and their meanings

I Olfactory

So called because it is the nerve of smell. (From olfacere (L): to smell.)

II Optic

So called because it is the nerve of sight. (From oculus (L): the eye.)

III Oculomotor

So called because it is the motor nerve to the extrinsic muscles of the eyeball, apart from the superior oblique and lateral rectus, also of the ciliary muscle and sphincter of the iris. (From oculus (L): the eye + movere (L): to move.)

IV Trochlear

So called because it supplies the superior oblique muscle which passes through the trochlea. (From trochilea (Gk)/ trochlea (L): a pulley.)

V Trigeminal

So called because it divides into three branches: ophthalmic, maxillary and mandibular. (From trigeminus (L): triplets or triple.)

VI Abducent

So called because it is the motor nerve of the lateral rectus of the eyeball and so causes the eye to turn to the side. (From Ab (L): away from + ducere (L): to lead.)

VII Facial

So called because its fibres fan out into the face. (From facies (L): a face.)

VIII Auditory

So called because it is the nerve of hearing. (From audire (L): to hear.)

also known as:
Vestibulo-cochlear

So called because nerves pass to the semicircular canals and cochlear. (From vestibulum (L): a passage + cochlea (L): a snail.)

IX Glossopharyngeal

So called because it passes to the tongue and pharynx. (From glossa (Gk): the tongue + pharyngx (Gk): the gullet.)

X Vagus

The wandering nerve. So called because the structures it supplies are less clearly confined spatially than those of the other cranial nerves. (From vagus (L): roaming or wandering.)

XI Accessory

So called because, in travelling with the vagus nerve for part of its course, it appears supplementary to it. (From accedere (L): to be added to.)

also known as:

Spinal Accessory

As above but with an indicator as to the extracranial part of its origin.

XII Hypoglossal

So called because it passes under the tongue. (From hypo (Gk): under + glossa (Gk): the tongue.)

Words pertaining to joint types

Amphiarthrosis
A secondary cartilaginous joint allowing limited movement either side of a central cartilaginous disc. (From amphi (Gk): on both sides + arthron (Gk): a joint.)

Arthrodia
A synovial joint permitting only gliding movements. (From arthron (Gk): a joint.)

Ball-and-socket
A synovial joint where a rounded bony head articulates with a socket-like cavity. Occurs at shoulder, hip and malleus/incus joints only.

Condyloid
A uni-axial synovial joint where an ovoid bony head (or knuckle) articulates with an ovoid cavity. (From kondylos (Gk): a knuckle + eidos (Gk): form.)

Diarthrosis
The general term for synovial joints. (From dia (Gk): through + arthron (Gk): a joint.)

Ellipsoid
A bi-axial synovial joint where an ovoid bony head articulates with an ovoid cavity. (From ellips (Gk): oval + eidos (Gk): form.)

Enarthrosis
(From en (Gk): in + arthron (Gk): a joint.) See ball-and-socket joint.

Ginglymus
(From ginglymos (Gk): a hinge.) See hinge joint.

Gomphosis
An immovable joint between a conical process and a socket e.g. the teeth. (From gomphos (Gk): a bolt.)

Hinge
A synovial joint allowing movement in one plane only.

Pivot
A synovial joint where movement is limited to rotation only.

Plane
A synovial joint between fairly flat articular surfaces.

Saddle
A bi-axial synovial joint where the opposing joints are concavoconvex i.e. concave

	in one plane and convex in the plane perpendicular to that.
Schindylesis	An immovable joint between a thin plate of bone and a cleft or fissure as occurs between the rostrum of the sphenoid and the vomer. (From schindylesis (Gk): a fissure.)
Sellar	(From sella (L): a saddle or seat.) See saddle joint.
Spheroidal	(From sphaira (Gk): a globe or sphere + eidos (Gk): form.) See ball-and-socket joint.
Sutures	An immovable fibrous joint between bones articulating by processes and interdigitations. (From sutura (L): a seam or a sewing together.)

Sutures were previously further subdivided as follows:

Sutura vera	(True sutures)
Dentata	Tooth-like sutures e.g. the sagittal suture.
Serrata	Sutures with serrated edges e.g. the (transient) metopic suture.
Limbosa	Sutures with bevelled edges e.g. the fronto-parietal suture
Sutura notha	(False sutures)
Squamosa	Sutures with thin bevelled edges e.g. the squamo-parietal suture
Harmonia	Sutures where contiguous roughened surfaces are in apposition e.g. the intermaxillary suture.
Symphysis	A joint with limited movement where the bones are connected by fibrocartilage. (From symphysis (Gk): growing together; from syn (Gk) together, with + phyein (Gk): to grow.)
Synarthrosis	An immovable joint where the bones are separated by a minimal amount of tissue. (From syn (Gk): together, with + arthron (Gk): a joint.)

Synchondrosis   An immovable joint where the bones are held together by cartilage. (From syn (Gk): together, with + chondros (Gk): cartilage.)

Syndesmosis   A partially movable joint where the bones are held firmly together by fibrous tissue. (From syndesmos (Gk): a ligament or band.)

Synostosis   The osseous union of two bones e.g. between the diaphysis and epiphysis of a long bone at the end of growth. (From syn (Gk): together, with + osteon (Gk): a bone.)

Synovial Joints   Readily movable joints. (From syn (Gk): together, with + ovum (Gk): egg. A term first used by Paracelsus (c. 1520 AD) apparently alluding to the nature of the synovial fluid.)

Trochoid   (From trochos (Gk): a wheel + eidos (Gk): form.) See pivot joint.

Words pertaining to action, movement and position in space

Abduction	To move a part *away from* the mid-line
Adduction	To move a part *towards* the mid-line
Afferent	Direction of flow or passage *towards* a reference structure
Anterior	Referring to the front
Axillary	Referring to the armpit
Bilateral	On both sides
Caudal	Towards the tail
Circumduction	A circular motion of a limb
Collateral	On each side
Contralateral	Referring to the opposite side
Cranial	Towards the head
Cubital	Referring to the (anterior) fossa of the elbow
Deep	Referring to a structure further from the surface than another
Dexter	The right side
Distal	Towards the free end of a limb
Dorsal	Pertaining to the back or towards the back
Dorsiflexion	To flex the ankle so as to turn the foot upwards
Efferent	Direction of flow or passage *away from* a reference structure
Eversion	To turn the plantar surface of the foot laterally
Extension	To increase the inner angle at a joint
Extensor	Sometimes used with reference to the posterior aspect of the arm (Not applicable to the leg)
External	Outside, usually with reference to a hollow structure
Fibular	Referring to the lateral aspect of the leg
Flexion	To decrease the inner angle at a joint

31

Flexor	Sometimes used with reference to the anterior aspect of the arm (Not applicable to the leg)
Hypaxial	Below or ventral to the vertebral column
Inferior	Towards the tail (caudal)
Inguinal	Referring to the groin
Internal	Inside, usually with reference to a hollow structure
Inversion	To turn the plantar surface of the foot medially
Ipsilateral	Referring to the same side
Lateral	Referring to the side
Lateral rotation	To turn the arm or leg, at the shoulder or hip, outwards with respect to the anatomical position
Longitudinal	Referring to the long axis of the body
Median	In the mid-line of the body
Medial	Referring to a structure relatively closer to the median plane than another
Medial rotation	To turn the arm or leg, at the shoulder or hip, inwards with respect to the anatomical position
Palmar	The surface of the hand directed forwards, in the anatomical position — the palm
Paramedian	On a plane parallel to that of the median (sagittal) plane
Parasagittal	On a plane parallel to that of the sagittal (median) plane
Peroneal	Fibular
Plantar	The surface of the foot upon the ground, in the anatomical position — the sole of the foot
Plantar flexion	To flex the ankle so as to turn the foot downwards
Popliteal	Referring to the (posterior) fossa of the knee
Postaxial	Posterior to an axis
Posterior	Referring to the back
Preaxial	Anterior to an axis

Pronation	To turn the palmar surface of the hand to face posteriorly
Proximal	Towards the trunk (fixed) end of a limb
Radial	Referring to the lateral aspect of the forearm
Rotate	To turn about a long axis
Sagittal	In an anteroposterior direction
Sinister	The left side
Superficial	Referring to a structure lying closer to the surface than another
Superior	Towards the head (cranial)
Supination	To turn the palmar surface of the hand to face anteriorly
Tibial	Referring to the medial aspect of the leg
Ulnar	Referring to the medial aspect of the forearm
Unilateral	On one side only
Ventral	Pertaining to the front or towards the front
Volar	Referring to the palmar aspect

Some anatomical lines, planes and points

For descriptive purposes, the human body is divided by imaginary lines and planes. The following is a list of such lines, planes and resulting points.

The Anatomical Position
A fully erect posture with feet side by side but slightly apart and arms a little abducted with the palms turned forward. All lines, planes and regions are described from this reference position.

Midaxillary Line
The line, in the coronal plane, passing through the middle of the axilla

Midclavicular Line
The line, parallel to the mid-line, passing through the middle of the clavicle

Mid-line
The line, in the median sagittal plane, dividing the body into two halves

Coronal Plane
A plane parallel to the coronal suture of the cranium

Horizontal Plane
A plane, parallel to the ground

Intertubercular Plane
The horizontal plane passing, at the level of the iliac crests, through the body of L5

Lateral Vertical Plane
The sagittal plane passing through the point midway between the anterior superior iliac spine and the mid-line

Louis' Plane
The horizontal plane passing through the angle of Louis (sternal angle) and the disc between T4 and T5

34

Median Vertical Plane	The median sagittal plane
Median Sagittal Plane	The mid-line plane dividing the body in two, running parallel with the sagittal suture of the cranium. Some suggest that the term 'median sagittal plane' is a misnomer and should simply be the 'median plane'.
Sagittal Plane	A plane parallel with the median sagittal plane, not in the mid-line
Spinous Plane	The horizontal plane passing through the anterior superior iliac spines. Passes just below the level of the sacral promontory
Sterno-xiphoid Plane	The horizontal plane passing through the junction of the sternum and xiphoid process and the disc between T9 and T10
Subcostal Plane	The horizontal plane at the lowest level of the ribs, corresponding to the level of L3
Supra-sternal Plane	The horizontal plane between the supra-sternal notch and the disc between T2 and T3
Thoracic Plane	The horizontal plane at the level of T7, posteriorly and the junction of the 4th costal cartilage and the sternum, anteriorly
Transpyloric Plane	The horizontal plane passing, midway between the supra-sternal notch and the symphysis pubis, through the pyloris (approx. L1)
Transtubercular Plane	See intertubercular plane
Transverse Plane	A plane parallel to the horizontal
Umbilical Plane	The horizontal plane passing between L3 and L4. (The supposed level of the umbilicus.)

Central Point   The point of crossing of the transpyloric and median sagittal planes

Lateral Central (or para-central) Point   The point of crossing of the transpyloric and lateral vertical plane

Angle of Louis   The sternal angle — the junction of the manubrium and body of the sternum

The regions of the abdomen

Epigastric Region
Bounded by the transpyloric plane, inferiorly and both lateral vertical planes, laterally

Hypochondriac Region (Left and Right)
Bounded by the transpyloric plane, inferiorly and either one of the lateral vertical planes, medially

Hypogastric Region
Bounded by the transtubercular plane, superiorly and both lateral vertical planes, laterally

Iliac Region (Left and Right)
Bounded by the transtubercular plane, superiorly and either one of the lateral vertical planes, medially

Lumbar Region (Left and Right)
Bounded by the transpyloric plane, superiorly, the transtubercular plane, inferiorly and either of the lateral vertical planes, medially

Umbilical Region
Bounded by the transpyloric plane, superiorly, the transtubercular plane, inferiorly and both lateral vertical planes, laterally

Named points on the skull

The use of named points on the skull was once more prevalent than it is today, especially within the related field of physical anthropology. The following list includes some words which although now infrequently used in anatomy are still to be encountered in some of the older yet still eminent texts.

The words follow an anatomical rather than an alphabetical order.

(i) Points in the median sagittal plane

Alveolar Point/Prosthion	The centre point, anteriorly, on the upper alveolar margin
Akanthion	The tip of the inferior nasal spine
Subnasal Point (or spinal point)	The mid-point on the inferior border of the base of the nasal spine
Rhinion	The most anterior point of the junction of the two nasal bones
Nasion	The mid-point of the naso-frontal suture
Glabella	The point midway between the two superciliary ridges
Ophryon (or Supraorbital point)	The mid-point of the narrowest transverse diameter of the forehead, as measured between the temporal lines. The transverse line through this point separates the face from the cranium
Metopion	The point midway between the frontal eminences
Bregma	The junction of the coronal and sagittal sutures
Vertex	The highest point of the cranium
Obelion	The region between the parietal foramina, where the sagittal suture is less complex than elsewhere
Lambda	The junction of the lambdoid and sagittal sutures

38

(Maximum) Occipital Point	The point in the median sagittal plane, over the occipital bone, that is furthest from the glabella
Inion	The external occipital protuberance
Opisthion	The mid-point of the posterior margin of the foramen magnum
Basion	The mid-point of the anterior margin of the foramen magnum
Pogonion	The most prominent point of the chin as evident on the mandible

(ii) Points on the side of the skull

Jugal Point	The angle between the vertical border and the temporal process of the zygomatic bone
Dacryon	The point, on the inner margin of the orbital cavity, where the vertical lacrimo-maxillary and fronto-nasal sutures meet
Pterion	In the region of the parieto-sphenoid suture. (Sutures in this region are variable incorporating varying proportions of the frontal and parietal bones and of the greater wing of the sphenoid and squamous part of the petrous bones.)
Stephanion	The point where the coronal suture and the temporal lines cross
Asterion	The point where the lambdoid, parieto-mastoid and occipito-mastoid sutures meet
Supra-Auricular Point	The point immediately superior to the middle of the external auditory meatus close to the edge of the root of the zygomatic process of the temporal bone
Gonion	The lateral side of the angle of the mandible

Head types

A popular measure of transverse head shape is the cephalic index. This index consists of expressing the maximum head width, between the parietal eminences, as a percentage of the head length, between the glabella and the occiput:

$$C.I. = \frac{\text{Head Width}}{\text{Head Length}} \times 100$$

The term *cephalic* index is used when this measure is taken from living subjects, whilst the term *cranial* index is used when it is taken from dry skulls.

Depending upon the index determined, the head is allocated into one of the following three groups:

Dolichocephalic Long-headed. C.I.<75. (From dolichos (Gk): long + kephale (Gk): the head.)

Mesocephalic Moderate-shaped head. C.I. between 75 and 80. (From mesos (Gk): middle + kephale (Gk): the head.)

Brachycephalic Short-headed. C.I.>80. (From brachys (Gk): short + kephale (Gk): the head.) The term 'hyper-brachycephalic' has also been used denoting a very short-headedness (C.I.>85) (From hyper (Gk): over, brachys (Gk): short + kephale (Gk): the head.)

Heads may also vary in height, being described, non-metrically as:

Acrocephalic A relatively pointed head. (From akros (Gk): summit + kephale (Gk): the head.)

Platycephalic A relatively flat head. (From platys (Gk): flat + kephale (Gk): the head.) Such a head may also be relatively wide.

Body types

Various methods for classifying physique have been attempted. Although metrical classifications have been generally unsuccessful, certain terms of a non-metrical nature persist.

Hypersthenic   A very muscular, thick-set individual, with a broad chest and high diaphragm. The stomach tends to lie transversely and the gall-bladder horizontal high in the abdomen, well away from the mid-line. The transverse colon is also high. (From hyper (Gk): over + sthenos (Gk): strength.)

Sthenic   A muscular, thick-set individual similar to hypersthenic but not quite as broad in relation to height. The stomach and gall-bladder lie more vertically and the transverse colon curves lower. (From sthenos (Gk): strength.)

Asthenic   A long, thin chested lean individual with the diaphragm, stomach, gall-bladder and transverse colon situated lower down than those types mentioned above. In the erect position the stomach and transverse colon descend into the pelvis. (From a (Gk): without + sthenos (Gk): strength.)

Hyposthenic   Individuals of this type are similar to those of the asthenic type, but the features mentioned are not as marked. (From hypo (Gk): under + sthenos (Gk): strength.)

Glossary

A

A	(Abbreviation) Artery (plur. Aa).
Abdomen	The belly. Of uncertain Latin etymology. May originally have been applied to the belly of pregnant pigs and in turn used as a term of ridicule.
Aberrans	An aberration; deviation from the normal or usual course. (From Ab (L): away from + errare (L): to stray.)
Aberrant	An aberration; deviation from the normal or usual course. (From Ab (L): away from + errare (L): to stray.)
Accessory	Accessory; supplementary. (From accedere (L): to be added to.)
Acetabulum	The concavity in the innominate bone forming the socket of the hip joint. (From acetum (L): vinegar + -abulum (dim. of -abrum) (L): a holder or receptacle. Literally, a small vinegar cup.) The acetabulum is named after the item of Roman tableware.
Achilles' Tendon	See Tendo Achilles.
Acinus	The smallest terminal secretory lobe of a compound gland. (From acinus (L): a grape or other juicy, seed-bearing berry.)
Acoustic	Pertaining to sound or the sense of hearing. (From akoustikos (Gk): pertaining to hearing.)
Acromion	A bony projection of the scapula. (From akros (Gk): summit + omo (Gk): shoulder. Literally, the highest point of the arm at the shoulder.) First used by Greek anatomists and adopted by Galen.

42

Adamantine   The enamel of teeth. (From adamas (L): diamond / a (Gk) not, without + damao (Gk): I tame. Literally, 'I cannot make to yield'.)

Adeno- **hypophysis**   The anterior lobe of the pituitary gland. (From aden (Gk): acorn, hypo (Gk): under + physis (Gk): growth.) Aden (in the singular) refers to glandular tissue; hypophysis, to the pituitary gland which appears to 'grow under' the brain.

Adipose   Fat. (From adeps (L): animal fat.) The word adiposus is not a Roman word but rather an invention from the end of the first millenium.

Aditus   An opening or an entrance. (From adire (L): to approach.)

Adnexa   Appendages or structures closely associated with an organ. (From ad (L): to + nectere (L): to bind.)

Adminiculum   A support. (From adminiculum (L): a support.)

Adrenal   The suprarenal glands. (From ad (L): to + rene (L): kidney. Literally, near to the kidney.)

Adventitia   Literally, a stranger, coming from abroad. (From ad (L):to + venire (L): to come). An adventitious structure is one that is in an unusual place. The tunica adventitia (the outer coat of a blood vessel) is a coat that is drawn from surrounding connective tissue.

Aequator   An equator, an imaginary line that divides a globe into hemispheres. (From aequator (Med L) from aequo (L): make equal.)

Affixus   Fixed. (From affigere (L): to affix.)

Agger   An embankment, eminence or mound. (From ager (L): a mound or eminence.)

Agonist   A muscle that is a prime mover. (From agonistes (Gk): an opponent in sport.)

Ala   A wing or wing-like structure. (From ala (L): a wing.) Properly, 'ala' referred to a mobile wing whereas 'pinna' referred to a stationary one. This is not followed strictly in anatomy.

Alba (Albus)   White. (From albus (L): white.)
                     Also as: albicans.

Albuginea   White or whitish. (From albus (L): white.)
                     This is not a Roman word but rather an
                     invention from the end of the first millenium.

Alimentary   Pertaining to the digestive tract. (From alere
                     (L): to nourish.)

Allantois   Literally, sausage-like. (From allas (Gk): a
                     sausage.) This term was used by early
                     anatomists because of the resemblance of this
                     structure which was first seen in the calf.

Alveolus   A small pit or cavity. (From alveolus (dim. of
                     alveus) (L): a small pit.) In Latin, it is used for
                     a variety of hollow objects. (Anatomically,
                     this term was first used by Vesalius with
                     reference to the tooth socket. Application to
                     the lung was made in 1846 by Rossignol
                     because of its resemblance to a honeycomb.)

Alveus   A trough or cavity. (From alveus (L): a
                     cavity.)

Ambiens   Surrounding. (From ambiens (L): going
                     around.)

Ambiguus   Doubtful. (From ambigere (L): to drive in
                     two directions.) Therefore of uncertain func-
                     tion, meaning etc.

Ammon's   A portion of the pes hippocampi resembling a
Horn   ram's horn. Named after Ammon, an Egyp-
                     tian god with a ram's head.

Amnion   The innermost membrane which surrounds
                     the fetus and contains the amniotic fluid.
                     (From amnos (Gk): lamb.) This term seems to
                     have derived from a knowledge of the fetal
                     membranes of the sheep, whether in the field
                     or from the sacrifice of pregnant sheep during
                     which these membranes would have been
                     exposed. (This term was first used in its
                     anatomical sense by Galen.)

Ampulla   A flask-shaped dilatation of a canal. (From
                     ampla (L): full + bulla (L): vase.) A Roman

ampulla was a clay or glass vase with a narrow neck.

Amygdala — A rounded lobe of the cerebellum. (From amygdale (Gk): an almond.)

Anastomosis — A communication. (From anastomosis (Gk): to provide with a mouth.)

Anatomical Snuff Box — The hollow formed, when the thumb is extended, between the tendons of extensor pollicis brevis and abductor pollicis longus anteriorly and extensor pollicis longus posteriorly on the radial side of the wrist.

Anatomy — The study of the form and structure of organisms, founded upon dissection. (From ana (Gk): apart + temnein (Gk): to cut. Literally, to cut apart.)

Angustia — A narrowing or constriction. (From angustus (L): narrow.)

Angle — An angle. (From angulus (L): an angle or corner.)

Angularis — Angular. (From angulus (L): an angle or corner.)

Ankle — Probably derived from angulus (L): an angle or corner.

Anlage — A precursory form or structure. (From an (Gm): on, near + legen (Gm): to lay or to place.)

Annulus — See Anulus.

Anomalous — A variant of a standard form. (From a (Gk): not + homalos (Gk): even.)

Ansa — A loop. (From ansa (L): the loop of a sandal or the handle of a bucket.)

Anserinus — Like a goose. (From anser (L): a goose.) See pes anserinus and cutis anserinus.
Also as: anserine.

Antagonist — A muscle opposing the pull of an agonist. (From anti (Gk): against + agonistes (Gk): an opponent.)

Ante-brachium — The forearm. (From ante (L): before + brachium (L): arm.)

Anteflexion   Bent forwards, the normal shape of the non-pregnant uterus. (From ante (L): before + flexere (L): to bend.)

Anteversion   Turned forwards, the normal position of the non-pregnant uterus. (From ante (L): before + vertere (L): to turn.)

Antihelix   The curvature of the pinna of the ear parallel to the helix. (From anti (Gk): before/against + helix (Gk): coil/convolution.)

Antitragus   Prominence of the pinna of the ear opposite the tragus. (From anti (Gk): before/against + tragus (Gk): he-goat.)

Antrum   A cavity. (From antron (Gk): A cave/ antrum (L) A cavity.)

Anulus   A small ring. The fourth (ring) finger may be so called. (Dim. of Anus (L): a ring.)

Anus   The exit of the alimentary tract. (From anus (L): a ring.) (This word may also have Sanskrit origins, from the word meaning 'to sit'.) (Anatomically, this term was first used by Celsus. This term was also used by the Romans to mean 'an old woman', the characteristic in common being the wrinkled skin.)

Aorta   The principal artery of the body. (From aorte (Gk): the great artery.)
The term may also come from aorteomai (Gk): 'I am suspended'. According to Aristotle, the aorta was the large artery from which the heart was suspended. (The term 'arch of the aorta' was first used in 1732 by Heister (anatomist and surgeon in Altdorf and Helmstadt).)

Apertura   An aperture or opening. (From aperire (L): to open.)

Apocrine   A secretory cell type where part of the protoplasm was believed to contribute to the secretion. (From apo (Gk): from + krinein (Gk): to separate.)

Apex
The peak, tip or top. (From apex (L): a summit.)
Also as: apical, pertaining to an apex.

Aponeurosis
A dense connective tissue sheet serving as a tendon. (From apo (Gk): from + neuron (Gk): tendon.

Aplasia
Incomplete or defective development of a structure. (From a (Gk): not + plastos (Gk): formed.)

Apophysis
A bony outgrowth or process of bone usually due to the pull of a muscle. (From apo (Gk): from + physis (Gk): growth.)

Appendix
An appendage. (From appendere (L): to hang on.)

Aqueduct
A canal or duct for the passage of fluid. (From aqua (L): water + ducere (L): to lead.)

Aqueous
Watery. (From aqua (L): water.)

Aqueous Humour
The watery substance of the eye. (From aqua (L): water + humour (L): liquid, bodily fluid.)

Arachnoid Granulations
The grain-like elevations of the arachnoid mater through the dura mater for the purpose of passing cerebrospinal fluid into the venous system. (From arachne (Gk) spider or cobweb, -eidos (Gk): like + granum (L): grain.)

Arbor Vitae
The tree-like white substance within the cerebellum. (From arbor (L): tree + vita (L): life. Literally, tree of life.) (Anatomically, this term was intoduced by Winslow. Properly, it refers to the Arbor Vitae Cerebelli since Winslow's term Arbor Vitae Uteri refers to folds within the canal of the cervix.)

Arcade
An arched channel or passage. (From arcus (L): a bow.)

Arch
An arch. (From arcus (L): an arch.)

Archicerebellum
The (evolutionarily) oldest or first part of the cerebellum. (From archi (Gk): first + cerebellum (dim. of cerebrum) (L): a small brain.)

Archipallium   Olfactory area of the cerebrum; the hippo-campus. (From archi (Gk): first + pallium (L): a mantle. Literally, the oldest or first layer of the cerebrum.)

Archi-striatum   The primitive corpus striatum. (From archi (Gk): first + striatum (L): furrowed or grooved.)

Arcus   An arch. (From arcus (L): an arch.)

Area   An area or surface. (From area (L): an area or surface.)

Areola   A small area. (Dim. of area (L): a space.) (It has been suggested that the term 'areola', as applied to the area around the nipple, is derived from aureolus (L): golden.)

Arm   From earm (OE): an arm.

Arrector   That which raises. (From arrigere (L): to raise.)

Arrector Pili   The raiser of a hair. A small smooth muscle bundle attached to the hair follicle that when contracted causes goose flesh (cutis anserina). (From arrigere (L): to raise + pilus (L): a hair.) (Pl. arrectores pilorum.)

Arteriole   A very small artery with a diameter of less than 200μm. (From arteriola (L): a small artery.)

Artery   A vessel conveying blood from the heart. (From Aer (Gk): air + terein (Gk): to keep; arteria (L): wind-pipe.) (It was believed for many centuries that arteries contained air and the veins, blood. Arterial pulsation was thought to be due to the action of the blood in the veins.)

Articulation   A joint. (From artus (L): a joint. Also articulatus (L): a little joint.)

Aspera   Rough. (From asper (L): rough.)

Astrocytes   Star-shaped neuroglial cells, also called mac-roglia. (From aster (Gk): star + kytos (Gk): cell.)

Atrium   A chamber of the heart or the dilated part of

the middle meatus of the nose. (From atrium (L): a hall or room.) (The atrium was the main room of the Roman house. Lacking windows, smoke from that room's central fire blackened its walls. Similarly, the walls of the atria of the heart are darkened by the post mortem action of blood.)

Atticus
The epitympanic recess. (From attikos (Gk) / Atticus (L): Athenian.) This term has been used to describe buildings of the Athenian style built one storey upon another.

Auditory
Pertaining to the ear. (From audire (L): to hear.)

Auricle
The external ear; a saccular appendage of the left atrium. (From auricula (dim. of auris) (L): ear.) (Auricle is also the obsolete term for an atrium per se. 'Auricularis' was the term applied by the Romans to the little finger which was used to poke the ear-hole.)

Autonomic (Nervous System)
The essentially self-regulating nervous system dealing with vegetative functions. (From auto (Gk): self + nomos (Gk): law.)

Axilla
The armpit (or axillary fossa). (From axilla (L): armpit.)

Axon
The process conducting action potentials away from the nerve cell body. (From axon (Gk): an axis.)

Azygos
The vein running on the right side from abdomen to thorax. (From a (Gk): not + zygos (Gk): yoked or paired. Literally, unpaired (due to its asymmetry).) This term was first used by Galen.

B

Basilar
Used to refer to something situated at the base of another structure. (From basis (L + Gk): a foundation or base.)

Basilic Vein
The basilic vein was so named by Avicenna at the end of the first millenium. It was a vein much used in blood-letting, particularly to cure a malady of the baser parts of the body which excluded the head. (From basilikos (Gk): important, prominent and Al-basilic (Ar): the inner vein.)

Basis
A base. (From basis (L and Gk): a base.)

Basisphenoid
The posterior part of the body of the sphenoid bone only distinguishable in the growing child. (From basis (Gk): base, sphen (Gk): wedge + -oeids (Gk): shaped. Literally, the base of the wedge-shaped (bone).)

Bicuspid
Having two cusps (valve flaps). (From bis (L): twice + cuspis (L): cusp.)

Bifid
Cleft or forked; divided into two. (From bis (L): twice + fidere (L): to cleave or split.)

Bigemina
Twofold. (From bigeminus (L): paired or double; from bi- (L): two + geminus (L): twin.)

Bile
A secretion of the liver. (From bilus (L): bile.) See choledochus.

Biliary
Pertaining to bile. (From bilus (L): bile.)

Biventer
Having two bellies. (From bis (L): twice + venter (L): a belly.)

Bladder
A membranous sac containing air or fluid. (From blaedre (OE): a bag.)

Blood
The vital fluid of the circulatory system. (From blod (OE): blood.)

Bone
The rigid framework of the body composed of connective tissue with a ground substance of mineral salts. (From bon (OE): bone.)

Brachium
The arm above the elbow. (From brachium (L): upper arm.)

Brain
The cranial centre of the nervous system. (From breagen (OE) and brayne (ME): brain.)

Branchial
Pertaining to the embryonic arches either side of the pharynx, posterior to the hyoid arch,

	which resemble gills. (From branchia (Gk): gills.)
Breast	The anterior chest wall. (From breost (OE): breast.)
Brevis	Short. (From brevis (L): short.)
Bronchiole	A small terminal branch of the bronchi. (Dim. of brongchos (Gk): windpipe.)
Bronchus	The passages from the trachea to each lung. (From brongchos (Gk): windpipe.) (Alternatively, this term may derive from brachein (Gk): to moisten. It was believed by Plato that the trachea was the passage for fluid drunk, whilst the oesophagus was the passage for food eaten.)
Buccal	Pertaining to the cheek. (From bucca (L): cheek.)
Bulbus	A bulb or swelling. (from bulbus (L): a bulb.)
Bulbus Oculi	The eyeball. (From bulbus (L): a bulb + oculus (L): the eye.)
Bulla	A rounded eminence. (From bulla (L): a water bubble or knob.)
Bursa	(1) A synovial membrane-lined sac interposed between two structures and containing synovial fluid. The use of this term in this meaning dates from the 17th century. (2) A peritoneal pouch. (From bursa (Gk): purse.)
Buttock	The gluteal region. (Dim. of butt (OE): end.)

C

C	The abbreviation pertaining to the vertebrae or nerves of the neck (Cervix). E.g. C5: the fifth cervical vertebra or nerve.
Cadaver	A dead body. (From cadaver (L): a corpse (from cadere (L): to fall).)
Caecum	The blind-ended sac of the proximal large intestine. (From caecus (L): blind.) (Anato-

mically, this term has also been applied, by Vesalius to the vermiform appendix.) (The Arabs referred to the caecum as the monoculum (the one-eyed) since it had one opening, that of the vermiform appendix.)

Caeruleus
Sky-blue, bluish. (From caeruleus (L): sky blue.)

Calamus Scriptorius
The area of the floor of the fourth ventricle, including the hypoglossal and vagal triangles, shaped like a pen nib. (From calamus (L): reed + scriptorius (L): a scribe. Literally, a writer's pen.) (This term was first used by Herophilus in the late 4th century BC.)

Calcaneus
The tarsal bone forming the heel. (From calx (L): heel.)

Calcar
A spur. (From calcar (L): a spur.)

Calcar Avis
An elevation, caused by the calcarine sulcus, in the posterior horn of the lateral ventricle, which resembles a bird's spur. (From calcar (L): a spur + avis (L): a bird.)

Calcification
The deposition of mineral salts in otherwise soft tissues. (From calx (L): lime + fadere (L): to make.)

Callosus
Hard. (From callum (L): hard skin, but may also mean a beam such as a rafter.)
Also as: callosum.

Callus
(1) An area of hard skin.
(2) An area of new bone formation following fracture.
(From callum (L): hard skin.)

Calvaria
The vault of the skull excluding the facial bones. (From calva (L): skull / calvus (L): bald.)

Calyx
A urine-collecting cavity of the kidney; subdivisions of the renal pelvis. (From kalyx (Gk): a cup or beaker.)

Camera
A chamber. (From camera (Gk): a chamber.)

Canal
A channel. (From canalis (L): a channel, passage or water pipe.)

Canaliculus   A small canal or channel. (Dim. of canalis (L): a channel, passage or water pipe.)

Canalis   A channel or base. (From canalis (L): a channel, passage or water pipe.)

Cancellus   A type of bone resembling a sponge-like framework. (From cancelli (L): lattice work.)

Canine   Dog-like. (From caninus (L): dog-like (from canis (L): a dog).)

Canthus   The junction of the upper and lower eyelids. (From kanthos (Gk): the metal rim around a wheel.) The present usage of this term derives from its former application, meaning the rim around both eyelids.

Capillary   A minute thin-walled blood or lymph vessel. (From capillus (L): a hair / capillaris (L): hair-like.)

Capillaris   Pertaining to the hair. (From pilus capitis (L): hair of the head.)

Capitulum   A small articular prominence at the end of a bone. (Dim. of caput (L): head.)

Capsula   A small box. (From capsula (dim. of capsa) (L): a little box.)

Capsule   An enclosing membrane of an organ. (From capsula (dim. of capsa) (L): a little box.)

Caput   A head. (From caput (L): a head.)

Cardia/
Cardiac   Pertaining to the heart. (From kardia (Gk): heart.)

Cardinal   A chief or principal structure. A hinge or pivot; that upon which something depends. (From cardo (L): a hinge.)

Carina   A keel-like ridge; the superiorly pointing ridge at the bifurcation of the trachea. (From carina (L): the keel of a boat.)

Carneae   Fleshy. (From carneus (L): meat.)

Carotid   The major arteries of the neck, supplying the head. (From karos (Gk): a heavy sleep.) It was believed from the time of Aristotle until the Middle Ages that compression of the carotid arteries led to the induction of a deep sleep or stupefaction. Thus, alternative names have included arteria soporariae, arteria soporiferae and arteria somni.

Cartilage A fairly flexible connective tissue composed of cells enclosed in a matrix of collagen, elastic fibre and mucopolysaccharides. (From cartilago (L): gristle, cartilage.) (First used, anatomically, by Celsus, in the early first century AD.)

Caruncle Any small fleshy eminence. (Dim. of caro (L): flesh.)

Cauda A tail or tail-like structure. (From cauda (L): a tail.)

Cauda Equina The spinal nerve roots within the neural canal of the lumbosacral vertebrae. (From cauda (L): a tail + equus (L): a horse. Literally, a horse's tail.)

Cavalry Bone An occasional ossification within the adductor longus tendon. Also known as rider's bone and exercise bone.

Caverna A cavity or space. (From caverna (L): a chamber or grotto.)

Cavernous Having spaces, sinuses or hollows. (From cavernosus (L): chambered.)

Cavity A space. (From cavitas (L): a cavity.)

Cavum A cavity or cave. (From cavum (L): a cavity or space.) See vena cava.

Cell A small cavity or hollow, a protoplasmic unit. (From cella (L): a compartment.)

Centrum (1) The central or middle point.
(2) The vertebral body exclusive of that part derived from the neural arch.
(From centrum (L): centre.)

Cephalic Pertaining to the head. (From kephale (Gk): head). The naming of the cephalic vein derives from the view that blood-letting from this vessel cured headache.

Cerebellum That part of the hindbrain lying inferior to the cerebrum and posterior to the fourth ventricle and pons. (Dim. of cerebrum (L): brain.)

Cerebrum	The brain. (From cerebrum (L): the brain.)
Cerumen	The waxy secretion from the ceruminous glands of the ear. (From cera (L): wax.)
Cervix	A neck. (From cervix (L): neck.)
Cervical	Pertaining to a neck. (From cervix (L): neck.)
Cheek	The wall of the mouth. (From ceoce (OE): cheek.)
Chest	Pertaining to the thorax. (From kiste (Gk): A chest or box; cista (L): a chest or a box.)
Chiasma	A crossing or decussation. Shaped like a cross or the Greek letter χ. (From kiasma (Gk): a cross.)
Choana	A funnel-shaped opening. (From choane (Gk): a funnel.)
Choledochus	The former name for the bile duct. (From chole (Gk): bile + dechomai (Gk): to receive.)
Chondral	Pertaining to cartilage. (From chondros (Gk): cartilage.)
Chondro-cranium	The skull when in a cartilaginous state. (From chondros (Gk): cartilage.)
Chorda	A cord or string-like structure. (From chorde (Gk): a string.)
Chorda Tympani	The tympanic cord or cord of the ear-drum. A branch of the facial nerve (VII) which passes through the tympanic cavity, medial to the tympanic membrane, to join the lingual branch of the mandibular nerve. (From chorde (Gk): a string + tympanum (Gk): a tambourine-like drum.) This structure was first described by Fallopius in the mid-16th century.
Chordae Tendineae	The tendinous cords running between the papillary muscles and atrioventricular valves within the heart. (From chorde (Gk): a string + tendere (L): to stretch.)
Chorion	An embryonic membrane outside and enveloping the amnion. (From chorion (Gk): skin.)

Chromosome The elements of the cell nucleus which stain darkly during mitosis and contain the genetic information. (From chroma (Gk): colour + soma (Gk): body.)

Chyle Lymph fluid with a high emulsified fat content. (From chylos (Gk): juice.)

Chyme Partially digested food having left the stomach. (From chymos (Gk): juice.)

Ciliary Pertaining to structures of the eyeball. (From cilium (L): an eyelash.)

Cilium The eyelash. (Originally the eyelid.) (From cilium (L): an eyelash.)

Cincereal Grey-coloured (or ashen). Therefore, pertaining to the grey matter. (From cinis (L): ash.) Also as: cincereum.

Cingulum A tract of cortical association fibres running from the frontal lobe to the tip of the temporal lobe. (From cingere (L): to girdle.) Also as: cinguli.

Circulation The movement of any fluid around a particular system. (From circulatio (L): the act of circulating.)

Circulus A circular or ring-like structure. (From circulus (L): a circle.)

Circumflex Bent around. (From circum (L): around + flexere (L): to bend.)

Cisterna An enclosed fluid-containing space. (From cisterna (L): a cistern or reservoir.)

Claustrum A thin strip of grey matter on the lateral surface of the external capsule, separating the lentiform nucleus from the insula. (From claustrum (L): a bar or barrier.)

Claustrum Vaginale The hymen. (From claustrum (L): a barrier + vagina (L): a sheath. Literally, the barrier to the vagina.)

Clava A small eminence on the posterior aspect of the medulla oblongata, containing the gracile nucleus. (From clava (L): a club.)

Cleido The Greek term for the clavicle. (From kleis (Gk): a key.)

Clitoris   The female erectile organ, homologous with the male penis, found at the anterior part of the vulva. (From kleiein (Gk): to enclose.) This term is derived from the suggestion that the vulva closes over the clitoris. Other suggestions are that the term is derived from kleis (Gk): a key and so refers to a doorkeeper; that it is derived from kleitorizein (Gk): to tickle, with obvious implications. The German term for the clitoris is der kitzler: the tickler!

Clivus   The slope on the sphenoid bone posterior to the sella turcica. (From clivus (L): a hill or slope.)

Cloaca   The chamber which is the common opening for discharge from the intestinal, urinary and genital tracts, in lower vertebrates e.g. birds and fish. This is also a transitory embryological feature in Man. (From cloaca (L): a drain or sewer.) In ancient Rome, the cloaca was a man-made canal carrying effluent to the Tiber.

Cluneal   Pertaining to the buttock. (From clunis (L): buttock.)

Cochlea   The conical spiral bony passage of the auditory system. (From cochlea (L): a snail and kochlias (Gk): a snail.) The term is derived from the resemblance to a snail shell.

Coeliac   Pertaining to the abdomen. (From koilia (Gk): the belly.)

Coelom   The intra- and extra-embryonic body cavities. (From koilos (Gk): a hollow or a cavity.)

Collagen   The fibrous protein making up the main organic component of connective tissue. (From kolla (Gk): glue + genos (Gk): offspring.) So called because when collagen-containing tissue is boiled, a glue or gelatin is produced.

Colliculus   A small hill. (Dim. of collis (L): a hill.)

Collum   A neck; the thinner part connecting an organ's head to its body or a bony head to the shaft. (From collum (L): neck.)

Colon   The large intestine. (From kolon (Gk): the large intestine.)

Columella   A small column. (Dim. of columna (L): a column.)

Columna   A column. (From columna (L): a column.)

Comes   A blood vessel running alongside another. (From comes (L): companion.)

Comitans   Accompanying. (From comitare (L): to accompany.)

Commissure   A connection or line of union. (From commissura (L): a joining together.) Applied (i) to neural tracts connecting bilateral centres or (ii) to fibrous bands joining two like structures.

Communi-   Communicating or connecting. (From com-
cans   municans (L): communicating.)

Conarium   The pineal body. (From conus (L): a cone / konarion (Gk): a little cone.) So called because of the resemblance with a cone.

Concha   A shell. (From concha (L): a shell / kongche (Gk): a shell.)

Condyle   A rounded, cartilage-covered bony eminence for articulation with another bone. (From kondylos (Gk): a knuckle.)

Confluence   A flowing together. (From confluere (L): running together.) Applied to the junction of superior sagittal and transverse venous sinuses.

Conjunctiva   The mucous membrane of the eye lining the eyelids, sclera and cornea. (From conjugere (L): to connect.) This term was not used by the Romans and is indeed not a Latin word at all. Instead, it was coined by Berengarius, around the turn of the 16th century, thinking that this membrane was continuous with the retina.

Connivens   Converging; arching over so as to meet. (From connivere (L): to wink.)

Conus   A cone. (From conus (L): a cone from konos (Gk): a cone.)

Conus Arteriosus   The infundibulum; the cone-shaped portion of the right ventricle from which the pulmonary trunk arises. (From conus (L): a cone + arteria (L): a wind pipe.)

Conus Medullaris   The tapering inferiormost portion of the spinal cord. (From conus (L): a cone + medulla (L): marrow.)

Convolution   A coiling or twisting. Applied to the ridges on the external surface of the brain. (From con (L): together + volvere (L): to wind.)

Copula   A ridge in the developing tongue. (From copula (L): bond.)

Cord   A cord-like structure. (From chorde (Gk): a cord.)

Corium   The (true) skin or dermis, below the epidermis. (From corium (L): skin, leather or hide.)

Cornea   The transparent tissue of the front of the eye. (From corneus (L): horny.)

Cornu   A horn. (From cornu (L): a horn.)

Corona   A crown. (From corona (L) a crown or a wreath.)

Coronary   Encircling, like a crown. (From corona (L): a crown or wreath.)

Coronary Arteries   The arteries of the heart. (From coronarius (L): pertaining to a crown or wreath.) So named because these vessels encompass the heart.

Corona Radiata   (1) The layer of cells surrounding the ovum after ejection from the ovary.
(2) The fibres running from the internal capsule to cerebral cortex.
(From corona (L): a crown + radiata (L): rays.)

Corpora   Bodies (From corpora (L): bodies (plur. or corpus).)

Corpora Cavernosa   The erectile tissue mass of the anterior wall of the penis. (From corpora (L): bodies + cavernosus (L): having cavities or spaces.)

Corpora Mamillaria   The mamillary bodies. (From corpora (L): bodies + mamilla (dim. of mamma) (L): a small breast.)

Corpora Quadri-gemina   The four rounded eminences forming the posterior part of the mesencephalon. (From corpora (L): bodies + quadrigeminus (L): fourfold.)

Corpus   A body. (From corpus (L): a body.)

Corpus Callosum   The strip of neural tissue joining the two cerebral hemispheres which is an area of tissue tougher than that adjacent to it and is a connecting beam. (From corpus (L): body + callum (L): hard.)

Corpus Luteum   The body found in the ovary after ejection of the ovum. (From corpus (L): body + luteus (L): yellow.)

Corpus Spongiosum   The erectile tissue mass forming the dorsal wall of the penis. (From corpus (L): body + spongia (Gk): a sponge.)

Corpus Striatum   The caudate and lentiform nuclei considered together. (From corpus (L): body + stria (L): a furrow, groove or flute in a column. Literally, the stripped body.)

Corpuscle   A little body. (From corpusculum (L): a little body.)

Corpusculum   A little body. (From corpusculum (L): a little body.)

Cortex   An outer layer or covering. (From cortex (L): the bark of a tree.)

Costa   Pertaining to a rib. (From costa (L): a rib.)

Cotyledon   A patch of villi within the placenta. (From kotyle (Gk): a cup.)

Coxa   The hip. (From coxa (L): the hip.) Also as: coxae meaning 'of the hip'.

Cranium   The skull. (From kranion (Gk): the skull.)

Cremaster   The muscular and connective tissue forming the intermediate layer of the spermatic cord.

(From kreman (Gk): to suspend.) So called because its contraction raises the related testicle.

Crista   A crest or ridge. (From crista (L): a crest, a cock's comb or tuft of hair on the head of an animal.)

Crista Galli   The crest-like ridge of the anterior cranial fossa on the superior part of the ethmoid bone. Literally, a cock's comb. (From crista (L): a crest, or cock's comb + gallus (L): cock.)

Crura   Legs or leg-like processes. (plur. of crus (L): a leg.)

Crus   A leg or leg-like process. (From crus (L): a leg.)

Crypt   A glandular tube or cavity. (From kryptos (Gk): hidden.)

Cubitus   The elbow. (From cubitus (L): elbow (also the distance between elbow and the tips of the fingers.) The term is derived from cubo (L): 'I lie down' or 'recline' (on one's arm), being the position which the Romans adopted during meals.

Cucullaris   A hood or hood-shaped. (From cucullus (L): a hood or cowl (like that of a monk.) Also a former name for the trapezius muscle, so-named because the two muscles together resemble a monk's cowl.

Culmen   The top or summit. (From culmen (L): top or summit.) Anatomically, it is the portion of the cerebellar vermis between the fissura prima and the postcentral fissure.

Cuneus   A wedge. (From cuneus (L): a wedge.) Applied to the wedge-shaped cortical area between parieto-occipital and calcarine sulci of the occipital lobe.

Cupola   A hole or hollow. (From kupe (Gk): a hole or a hollow.)

Cupula   The bony apex of the cochlea; the apex of the lung (From cupa (L): a tub or cup.)

Cusp   A prominence, pointed elevation. (From cuspis (L): a point; also a javelin or spear.) Applied to the points of the teeth.

Custodes   The defenders of virginity. (From custos (L):
Virginitatis defender, guard or keeper + virginitas (L): virginity or maidenhood.) A name applied to the adductor muscle group of the thigh (or by some to the Gracilis M. alone) because of their action in bringing the knees together.

Cutaneous   Pertaining to the skin. (From cutis (L): skin.)

Cuticle   The superficial skin layer. (Dim. of cutis (L): skin.)

Cutis   Skin. (From cutis (L): skin.)

Cutis   Goose flesh. (From cutis (L): skin + anser
Anserina   (L): a goose.)

Cyclops Fetus A developmental abnormality where one central rather than two normal eyes form.

Cymba   A hollow structure; a cup; a boat (From cymba (L) / kymbe (Gk): a boat.)

Cymba   That part of the external ear lying above the
Conchae   crus helicis. (From cymba (L) / kymbe (Gk): a boat + concha (L): a shell.)

Cyst   A bladder or sac-like structure. (From kystis (Gk): a bladder or pouch.)

Cystic   Pertaining to a bladder. (From kystis (Gk): a bladder or pouch.)

D

D   The abbreviation sometimes used to pertain to the dorsal (or thoracic) vertebrae. E.g. D5: the fifth dorsal vertebra.

Dacrocyst   The lacrimal sac. (From dakryon (Gk): a tear.)

Dactyl   A finger or digit. (From daktylos (Gk): a finger.)

Dartos   The thin layer of unstriated muscle at the base of the scrotum. (From dartos (Gk): flayed or skinned.) This term was first used at the turn of the first century by Rufus referring to the covering of the testes.

Decidua The mucous membrane of the uterus shed after childbirth. (From de (L): away + cadere (L): to fall.)

Deciduous Applied to the teeth which normally fall out. (From de (L): away + cadere (L): to fall.)

Declive A slope. (From declive (L): a hill.)

Decussation A crossing over. (From decussare (L): to cross over.)
Also as: decussatio.

Defecate To discharge faeces. (From defaecare (L): to cleanse.)

Deferens To carry down. (From deferre (L): to carry down.) The vas deferens were named, wrongly, by Berengarius of Carpi (1470–1530) who thought that sperm passes downwards instead of upwards.

Dehiscence The opening of an organ or structure along certain lines or in a specific direction. (From de (L): away + hiscere (L): to gape.)

Demilunar Half-moon-shaped. (From demi (L): half + luna (L): the moon.)

Dendron The branched process of a nerve cell reminiscent of a tree. (From dendron (Gk): a tree.)

Dendrite Pertaining to a dendron. (From dendron (Gk): a tree.)

Dens A peg-like process. (From dens (L): a tooth.)

Dentine The principal apatite-containing substance of the tooth which surrounds the pulp. (From dens (L): a tooth.)

Derm The integumental layers beneath the epidermis. (From derma (Gk): skin.)

Dermatome A definable area of skin innervated by the cutaneous branch of a specific spinal nerve. (From derma (Gk): skin + temnein (Gk): to cut up.)

Dermato-glyphics The ridges on the digits that produce fingerprints. From derma (Gk): skin + glypho

(Gk): I carve. Literally, skin-carvings.) This term was first used by Sir Francis Galton in 1892.

Descendens　Descending. (From descendere (L): to descend.)

Descensus　Descent. (From descendere (L): to descend.)

Diaphragm　The musculotendinous partition separating thorax and abdomen. (From diaphragma (Gk): partition or fence.) The use of this term to denote the structure now so named originates from Galen in the late second century AD.

Diaphysis　The shaft of a long bone. (From diaphysis (Gk): growing through.) This term derives from the shaft being situated between the two growing (epiphyseal) ends.

Diencephalon　That part of the brain forming the walls of the third ventricle. (From dia (Gk) through / between + encephalos (Gk): the brain.)

Digastric　Two-bellied. (From dis (Gk): double + gaster Gk): a belly.)

Digestive tract　The alimentary tract. (From digerere (L): to digest, to distribute.)

Digit　A finger or toe. (From digitus (L): a finger or toe.)

Dilatation　An expansion of or within an organ. (From dilatare (L): to expand.)

Diploë　The osseous tissue situated between the inner and outer compact lamellae of the flat cranial bones. (From diploos (Gk): double.)

Disc　A disc- or plate-like structure. (From discus (Gk): a disc.)

Diverticulum　A blind-ended recess or sacculation of a hollow structure. (From divertere (L): to turn aside.) This term is derived from the Roman term meaning 'to leave a road or passage'.

Duct　A tube-like passage for the passage of fluids; occasionally including blood or air. (From ducere (L): to lead.) Also as: ductus.

Ductule	A small duct. (Dim. ductus (L): a duct.)
Ductulus	A small duct. (Dim. ductus (L): a duct.)
Duodenum	The portion of the small intestine between the pyloris of the stomach and the jejunum. (From duodeni (L): twelve each.) This term derived from Herophilus' description, in the mid-fourth century BC, of this structure as being twelve fingers breadth in length. (The Greeks called the duodenum 'ekphysis', meaning 'a growing out' (of the stomach) since they thought it diverticular in nature.)
Duplicitus	Double or duplicated. (From duplicare (L): to fold in two.)
Dura	Hard. Applied to the dura mater, the tough outermost meningeal covering of the brain. (From durus (L): hard.) The term originates from the translation by Stephen of Antioch of Hali Abbas (9th century).

E

Ear	The organ of hearing. (From eare (OE): the ear.)
Ebur	Ivory or like ivory. (From ebur (L): ivory.)
Ebur Dentis	Tooth enamel. (From ebur (L): ivory + dens (L): a tooth.)
Eburnea	An old term for tooth enamel. (From ebur (L): ivory.)
Ectoderm	The outermost germ layer of the embryo. (From ektos (Gk): outside + derma (Gk): skin.)
Ectomeninx	The dura mater. (From ektos (Gk): outside + meningx (Gk): a membrane.) Also called the pachymeninx.
Ectopia	The condition when an organ is found in other than its usual place. (From ektopos (Gk): removed, away from a place, distant.)

Ejaculatory   Throwing out. (From ex (L): out + jacere (L): to throw.)

Elastic   Having the property of being able to return to the original form after deformation. (From elastikos (Gk): impulsive.)

Elastin   The component of elastic tissue. (From elaunein (Gk): to drive / elastikos (Gk): impulsive.) The term derives from the way in which the tissue returns itself to its prestretched form.

Elbow   From elboga (OE): elbow. From el- (OE): L-shaped + boga (OE): a bending or bow.

Embryo   The organism at an early antenatal stage in its intrauterine development. (From embryon (Gk): embryo + bryein (Gk): to grow or swell.)

Eminence   A protuberance or jutting out. (From eminere (L): to stand out or project.)

Eminentia   A protuberance or eminence. (From eminere (L): to stand out or project.)

Emissary   A coming out. (From emittere (L): to send out.)

Enamel   The hard material forming the outer cap of the tooth. (From esmaillier (OF): to coat with enamel / smatto (LL): I smelt.)

Encephalon   The brain. (From enkephalos (Gk): the brain. From en (Gk): within + kephalos (Gk): the head.)

End artery   A small branch of an artery, being the principal supply of a volume of tissue. Obstruction of such a vessel results in the death of the tissue supplied. This term was introduced by Cohnheim (1839–1884).

Endo-   Prefix: Inside or within. (From endon (Gk): within.)

  -cardium   Inside the heart. (From kardia (Gk): heart.)
  -chondral   Within cartilage. Applied to a form of ossification. (From chondros (Gk): cartilage.)
  -cranium   Inside the cranial cavity. (From kranion (Gk): the skull.)

-crine Organs which secrete ductlessly directly into the blood stream. (From krinein (Gk): to separate.)

-derm The embryonic epithelium of the respiratory system and the digestive system and associated organs. (From derma (Gk): skin.)

-lymph The fluid of the membranous labyrinth of the inner ear. (From lympha (L): water.)

-meninx A combined term for the pia and arachnoid maters. (From meningx (Gk): a membrane.) Also called the leptomeninx.

-metrium The mucous membrane lining of the uterus. (From metra (Gk): the uterus.)

-mysium The connective tissue separating muscle fibres. (From mys (Gk): a muscle.)

-neurium The connective tissue holding together and supporting nerve fibres. (From neuron (Gk): a nerve.)

-skeleton A skeleton inside the body. (From skeletos (Gk): hard.)

-steum The tissue lining the medullary cavity of a bone. (From osteon (Gk): a bone.)

-thelium The squamous epithelium lining of the heart, blood vessels etc. (From thele (Gk): nipple.) Probably a term referring to the sensitivity of the nipple and the delicacy of the lining.

Enteric Pertaining to the alimentary tract. (From enteron (Gk): the intestine.)

Eparterialis Situated above an artery. (From epi (Gk): upon, plus artery (see Artery.) Opposite of hyparterialis.

Ependyma The membranous lining of the cavities of the brain and the central canal of spinal cord. (From epi (Gk): upon + endyma (Gk): a garment. Literally, a clothing upon.)

Epi- Prefix: Upon. (From epi (Gk): upon.)

-canthus The elongation of skin over the inner angle of the upper eyelid. (From kanthos (Gk): corner.)

-cardium   The visceral layer of the pericardium. (From kardia (Gk): heart.)

-condyle   A bony protuberance above a condyle of a bone which does not participate in articulation. (From kondylos (Gk): a knuckle or knob.)

-cranium   Pertaining to the cranium. (From kranion (Gk): skull.)

-dermis   The most external layer of the skin. (From derma (Gk): skin.)

-didymus   That structure on the posterior and upper surfaces of the testicle. (From didymos (Gk): originally meaning a twin or paired, later meaning a testicle.)

-glottis   The leaf-shaped piece of elastic cartilage situated just superior to the entrance of the larynx. (From glottis (Gk): the larynx.)

-hyal   Applied to the upper surface of the hyoid bone. The epihyal bone is the ossified stylohyoid ligament. (From hyoeides (Gk): U-shaped.)

-mysium   The sheath enclosing a whole muscle. (From mys (Gk): a muscle.)

-nephros   The suprarenal or adrenal gland. (From nephros (Gk): kidney.)

-neurium   The sheath enclosing a peripheral nerve. (From neuron (Gk): a nerve.)

-physis   A secondary centre of ossification at the end of a long bone. Between it and the shaft of the long bone is situated the epiphyseal growth plate. (From phyein (Gk): to grow.)

Epiploic   The omentum. (From epiploon (Gk): a net, the caul of the entrails.)

(Epi-)

-thalamus   The area comprising the habenula, pineal body and posterior commissure of the brain. (From thalamos (Gk): a chamber.)

-thelium   Any cellular layer lining the free surface of a tube, cavity etc. (From thele (Gk): nipple.)

Initially, this term was used to refer to the skin of the lips, presumably echoing the sensitivity of the nipple and the delicacy of the lining.

-trichium The external layer of the epidermis of the fetus. (From thrix (Gk): hair.)

-tympanic Above the tympanic membrane. (From tympanum, (L): a drum.)

Eponychium The cuticular fold overlaying the lunula of the nail. (From onyx (Gk): a nail.)

Exitus An outlet. (From exite (L): to go out.)

Exocrine Applied to glands that drain by ducts. (From exo- without + krinein (Gk): to separate.)

Exteroceptor A neural receptor receiving stimuli from outside the body. (From externus (L): outside + capere (L): to take.)

Extra-pyramidal Outside the pyramidal system. The system concerned with muscle movement but not controlled at will. (From extra (L): outside + pyramis (L): a pyramid.)

Extrauterine Outside the uterus. Applied to life after birth. (From extra (L): outside + uterus (L): womb.)

Extremity A limb. (From extremitas (L): terminal portion, an extremity.)

Extrinsic A structure outside another to which it is related. (From extrinsecus (L): on the outside.)

Eye The organ of sight. (From ighe (ME) / eage (OE): the eye.)

Eye-tooth A term applied to the upper third or canine tooth. It possesses the longest root of all the teeth, directed towards the eye, and was believed by Galen to receive a nerve from the nerve supplying the eye.

Exercise Bone An occasional ossification within the adductor longus tendon. Also known as cavalry bone and rider's bone.

F

Fabella	A small sesamoid bone sometimes found in the lateral head of the gastrocnemius muscle. (From fabella (dim. of faba) (L): a small bean.)
Facet	A smooth round or flat surface on a bone usually associated with articulation. (From facette (Fr): a face.)
Facial	Pertaining to the face. (From facies (L): a face.)
Facies	A face or outer surface. (From facies (L): a face or outer surface.)
Faeces	The excrement from the bowel. (From faeces (plur. of faex) (L): refuse or dregs.)
Falx	A sickle-shaped fold in the dura mater. (From falx (L): a sickle.)
Fascia	A connective tissue sheet investing and separating individual muscles and muscle groups. (From fascia (L): a band or bandage, swaddling clothes.)
Fasciculus	A small bundle or cluster of fibres. The term is usually applied to muscles or nerves. (From fasciculus (dim. of fascis) (L): a bundle or packet.)
Fasciola	A narrow band. (From fasciola (dim. of fascia) (L): a small band or bandage.) Also as: fasciolaris.
Fastigius	Having a peak or summit. (From fastigium (L): a peak or summit.)
Fauces	The upper portion of the throat between the palate and the pharynx. (From faux (L): the throat or narrow passage.)
Faveolus	A small pit or depression. (From faveolus (dim. of favus) (L): a small pit or small honeycomb.)
Fel	Bile. (From fel (L): bile.)
Felleus	Pertaining to bile. (From fel (L): bile.)
Fenestra	A window-like opening. (From fenestra (L): a window.)

Ferruginea   Rust coloured. (From ferrugineus (L): rust-coloured.)

Fetus   The organism from the start of the third month of its intrauterine development. (From fetus (L): offspring.)

Fibre   A fibre-like structure. (From fibra (L): a thread or fibre.) In its present usage, this term dates from the time of Vesalius (mid-16th century).

Fibril   A small thread-like structure. (From fibrilla (L): a small fibre.)

Fibro-cartilage   A cartilage type containing visible collagenous fibres. (From fibra (L): a thread or fibre + cartilago (L): gristle or cartilage.)

Filament   A thin thread-like structure. (From filamentum (L): a thin thread.)

Filum   A thread-like structure. (From filum (L): a thread.)

Fimbria   A fringe-like structure. (From fimbria (L): a fringe.)

Finger   A digit of the upper limb. (From a teutonic root fingro- via Anglo-Saxon, but may come from an earlier root meaning - five.)

Fissure   A cleft or groove dividing an organ into lobes. (From fissura (L): a cleft or fissure.)

Fistula   An abnormal or anomalous communication between hollow structures with the outside or with each other. (From fistula (L): a little pipe.)

Flavum   Yellowish. (From flavus (L): yellow.)

Flexure   A bend. (From flexura (L): a bend.)

Flocculus   The small lobule of the posterior lobe of the cerebellum. (From flocculus (dim. of floccus) (L): a small tuft.)

Foetus   Alternative spelling of fetus. (From fetus (L): offspring.)

Folia   A leaf-like structure or formation. (From folium (L): a leaf.)

Follicle
A small sac or bag-like structure. (From folliculus (dim. of follis) (L): a small bag.)

Folliculus
A small sac or bag-like structure. (From folliculus (dim. of follis) (L): a small bag.)

Fontanelle
The unossified menbranous areas between the bones of the skull evident, depending upon the fontanelle in question, up to eighteen months after birth. (From fontanelle (Fr): a little fountain / fons (L): a fountain, spring or source.) The term may have derived from a procedure performed in the Middle Ages in which the skull was opened at the junction of the coronal and sagittal sutures to supposedly allow out the 'poison', causing a cerebral or visual condition.

Fonticulus
A small spring or fountain (see fontanelle). (From fons (L): a fountain, spring or source.)

Foot
The terminal portion of the lower limb. (From fot (OE): a foot.)

Foramen
An opening or hole. (From foramen (L): an opening / forare (L): to pierce.)

Formation
An arrangement or formation. (From formare (L): to form.)

Fornix
A structure resembling a vault or its arch. (From fornix (L): a vault or cellar or the arch of a vault.)

Fossa
A pit or depression in a structure. (From fossa (L): a pit, ditch or depression.)

Fossula
A small fossa. (Dim. of fossa (L): a pit, ditch or depression.)

Fourchette
A fold of mucous membrane at the posterior junction of the labia majora. (From fourchette (Fr): a fork.)

Fovea
A pit or fossa. (From fovea (L): a pit or grave.)

Foveola
A small fovea. (Dim. of fovea (L): a pit or grave.)

Fraenum
A bridge or ligament. (From fraenum (L): a bridge or ligament.)

Frenulum	A fold joining two structures. (From frenulum (dim. of frenum) (L): a small bridle.)
Frons	The forehead. (From frons (L): the forehead.)
Frontal	Pertaining to the forehead. (From frons (L): the forehead.)
Fulcrum	A support or a post. (From fulcire (L): to prop up or support.)
Fundus	A base or bottom. (From fundus (L): a base or bottom.)
Funiculus	A small cord, rope or bundle. (From funiculus (dim. of funis) (L): a small cord, rope or bundle.)

G

Galea	A helmet. (From Galea (L): a helmet.) Applied to the epicranial aponeurosis of the occipito-frontalis M. This term was first used by Santorini at the turn of the 18th century.
Gall	Bile, a secretion of the liver. (From gealla (OE): gall.)
Gallus	A cock. (From gallus (L): a cock.)
Ganglion	A group of nerve cells outside the central nervous system. (From ganglion (Gk): a knot, swelling or tumour.) Although originally coined to indicate a pathological condition, Galen (2nd century AD) applied this term to the sympathetic ganglia.
Gastric	Pertaining to the stomach. (From gaster (Gk): a belly or the stomach.)
Gastrula	An early stage in embryological development when parts of the blastula invaginate. (Dim. of gaster (Gk), thus little belly.)
Gelatinosa	Jelly-like. (From gelare (L): to congeal.)
Genial	Pertaining to the chin. (From geneion (Gk): the chin.)
Genital	Pertaining to the reproductive organs. (From gignere (L): to beget.)

Genu A knee. (From genu (L): a knee.)

Geniculatum Geniculate; bent like a knee. (From geniculum (L): a little knee.)

Geniculum A little knee. (From geniculum (L): a little knee.)

Gingiva The connective tissue covering of the alveolar margins of both jaws. (From gingivae (L): the gums.)

Girdle A belt, zone or supporting structure. (From gyrdan (OE): to gird.)

Gladiolus A small sword. A term sometimes applied to the body of the sternum. (From gladiolus (dim. of gladius) (L): a small sword.)

Gland A cellular arrangement for the elaboration of a secretion. (From glandula (dim. of glans) (L): a small acorn.)

Glans The distal part of the penis or clitoris. (From glans (L): an acorn.)

Glia The connective tissue of the central nervous system. (From gloia (Gk): glue.)

Globus A sphere, globe or ball. (From globus (L): a sphere or globe.)

Globus Pallidus The pale coloured medial portion of the lentiform nucleus. (From globus (L): a sphere or globe + pallidus (L): pale.)

Glomerulus The tuft of capillaries in the renal corpuscle. (From glomerare (L): to roll up.) This structure was first described by Malpighi in the late 18th century

Glomus A tuft of blood vessels. (From glomus (L): a ball.)

Glossal Pertaining to the tongue. (From glossa (Gk): the tongue.)

Glottis The narrow part of the larynx at the level of the vocal folds. (From glottis (Gk): the larynx.)

Gluteal Pertaining to the buttock. (From gloutos (Gk): the rump or buttock.)

Gonad The sexual organ producing the germ cells. (From gone (Gk): birth or generation.)

Gracile   Slender. (From gracilis (L): slender.)

Granulation   A grain-like structure. (From granum (L): grain.)

Granulosum   Grain-like. (From granum (L): grain.)

Griseum   Bluey-grey. (From griseus (L): bluish or grey.)

Groin   The furrow between abdomen and thigh. (From grein (Icel): branch.)

Groove   Any furrow or elongated depression. (From groef (Dut): channel.)

Gubern-aculum   The fibro-muscular cord, running from the epididymis to the scrotal wall, which directs the testes in its descent. (From gubernare (L): to govern.)

Gums   The tissues of the mouth related to the jaws. (From goma (OE): the jaw.)

Gustatory   Pertaining to the sense of taste. (From gustare (L): to taste.)

Gut   The intestine. (From gut (OE): a channel.)

Gyrus   A cerebral convolution. (From gyrus (L): a circle, ring or convolution.)

H

Habenula   A whip- or strap-like structure. (From habenula (dim. of habena) (L): a small strap or whip.)

Haemor-rhoidal   A term applied anatomically to the blood vessels of the anus and rectum. (From haima (Gk) blood + rhoia (Gk): to flow. Literally, likely to bleed.)

Hallux   The great toe. (From hallux (Mod L): the great toe.) This term is not found in classical Greek or Latin. Isidorus (circa 600 AD) uses the word 'allex' referring to the thumb or the great toe.

Hamstrings   The tendons of the knee flexors (gracilis, sartorius, semitendinosus, semimembranosus and biceps femoris). (From hamm (OE) the back of the thigh.)

Hamulus A hook-shaped process. (From hamulus (dim. of hamus) (L): a small hook.)

Harmonia A term that has been applied to the smooth articulation of bones over smooth joint surfaces. (From harmonia (Gk): harmony.)

Haustrum A sacculation of the wall of the large intestine (the colon). (From haustra (L): a sacculation or a pouch.)

Head From heafud or heafd (OE): the head.

Heart The pump for the circulatory system. (From heorte (OE): the heart.)

Helicotrema The point of communication, at the apex of the cochlea, between the scalae tympani and vestibuli. (From helix (Gk): a spiral + trema (Gk): a hole.)

Helix The circumferential, in-rolled portion of the pinna of the ear. (From helix (Gk): a spiral.)

Hepar The liver. (From hepar (L): the liver.)

Herniation A protrusion of a structure through its covering. (From hernia (L): a protrusion.)

Hiatus An opening or a cleft. (From hiatus (L): a gap.)

Hilum The defined region where nerves and vessels enter and leave an organ. (From hilum (L): a little thing, a trifle.)

Hilus The defined region where nerves and vessels enter and leave an organ. (From hilum (L): a little thing, a trifle.)

Hind brain The rhombencephalon.

Hip (1) The ball-and-socket joint of the femur and os innominatum.

(2) The os innominatum. (From hype or hiop (OE); hipe (ME): a bend or joint.)

Hippocampus The invaginated part of the cortex of the telencephalon. (From hippocampos (Gk): a sea-monster in Greek mythology.)

Holocrine A form of secretion in which the cell breaks down and becomes a part of the secretion. (From holos (Gk): whole + krinein (Gk): to separate.)

Horn A horn-like projection. (From horn (OE): a horn.)

Humour The fluid of the eye. A bodily liquid or fluid. (From humor (L): a bodily liquid or fluid.)

Hyaline The shiny bluish cartilage of joints. (From hyalos (Gk): glass.)

Hymen The membrane of the lower vagina, partially closing the entrance in the virgin state. (From hymen (Gk): a membrane.) Originally applied to any membrane, its current usage dates from Vesalius (mid-16th century). Hymen was the Greek god of marriage.

Hyparterialis Situated below an artery. (From hypo (Gk): below, plus artery (see Artery).) Opposite of eparterialis.

Hypophysis The pituitary body. (From hypo (Gk): below + physis (Gk): growth.) So called because the pituitary body appears to grow from beneath the brain.

Hypo-thalamus The part of the brain below the thalamus, forming the floor and side walls of the diencephalon. (From hypo (Gk): below + thalamus (Gk): a room.)

Hypothenar The palmar eminence of the medial side. (From hypo (Gk): below + thenar (Gk): the palm of the hand.)

Hypo-tympanum The lowest part of the tympanic cavity. (From hypo (Gk): below + tympanum (Gk): a drum.)

I

Ileum The distal portion of the small intestine. (From eilein (Gk): to twist.) This term was used by Galen (2nd century AD) but only in a pathological sense. The term's application to the structure in question dates from the Introductio Anatomica of 1618 (author unknown).

Ima Lowest. (From ima (L): lowest.)

Imbrication The overlapping of tile-like structures, such as the laminae of the thoracic spine. (From imbricare (L): to tile.)

Impar Unpaired or unequal. (From impar (L): without a fellow, unequal, odd.)

Impression A groove or shallowing. (From imprimere (L): to press in.)

Incertus Indefinite or undetermined. (From incertus (L): indefinite, undetermined or doubtful.)

Incisor A cutting tooth. (From incidere (L): to cut into, to notch.)

Incisura A notch (as if cut into). (From incidere (L): to make a cut into, to notch.)

Index (1) The forefinger or second digit. (From index (L): a pointer.)
(2) An index may also be a mathematical value denoting a relationship between two anthropometric values.

Indusium A layer. (From indusium (L): a tunic.)

Infundibulum A funnel-shaped structure. (From infundibulum (L): a funnel / infundere (L): to fill.)

Inguinal Pertaining to the groin. (From inguen (L): the groin.)

Innominate Without a name. (From in (L): not + nomen (L): name.)

Innominate artery So named because Galen (2nd century AD) described it without applying a name.

Innominate bone So named because Vesalius (mid-16th century) referred to the 'hip-bone' as 'unnamed'.

Insertion The attachment of a muscle which is the more mobile and usually the more distally or laterally situated. (From in (L): into + serere (L): to plant.)

Insula	That area of cerebral cortex which lies deep in the lateral fissure of the temporal lobe. (From insula (L): an island.)
Integument	The covering of the body, the skin. (From in (L): on + tegmen (L): a roof.)
Intercalatus	Being placed in between. (From intercalaris (L): inserted.)
Interdigit-ation	A finger-like interlocking of processes. (From inter (L): between + digitus (L): a digit or finger.)
Interstitial	The intervening spaces or areas within a tissue or an organ. (From inter (L): between + sistere (L): to set.)
Intestine	The gut. (From intestina (L): the entrails or gut.) This term is first found in Celsus (early 1st century AD).
Intima	The innermost layer in the wall of a blood vessel. (From intimus (L): innermost.)
Intrinsic	Pertaining to a structure situated within an organ. (From intrinsecus (L): situated on the inside.)
Introitus	An orifice or point of entry. (From introitus (L): entrance.)
Intumescence	An enlargement or swelling. (From intumescere (L): to swell.)
Invagination	To draw into a sheath. (From in (L): into + vagina (L): sheath.)
Involution	A deformation of a structure due to less favourable conditions. (From involutus (L): rolled up.)
Iris	The pigmented membrane of the eye behind the cornea forming the boundary of the pupil. (From iris (Gk): a rainbow.)
Islet	A small island. (Form insula (L): an island.)
Isthmus	A narrow connecting band or strip. (From isthmos (Gk): a connecting band or neck.)
Iter	A passage or canal; an aqueduct. (From iter (L): a way or journey.)

J

Jejunum	The proximal part of the small intestine. (From jejunus (L): empty, famished.) Since this part of the intestine seems emptier than the rest, Galen applied the Greek term 'nestis' (fasting) to this part. Jejunus is its Latin equivalent.
Jugular	Pertaining to the neck. (From jugulum (L): the hollow of the neck above the clavicle; the collar-bone.)
Jugum	A portion that unites. (From jugum (L): a yoke or collar.)
Junctura	A joint or articulation. (From junctura (L): a joining or articulation.)

K

Keratin	The basic protein substance of the horny layer of the skin, hair, nails etc. (From keras (Gk): horn.)
Kidney	The paired organ that excretes urine. (From cwith (OE): womb or belly + neere (OE): kidney.)
Knee	The joint between femur and tibia. (From cneow (OE): knee.)
Knuckle	The joints of the fingers. (From cnuel (OE) and knokll (ME): a bone.)
Kyphosis	The bend of the thoracic part of the spine; more properly, a pathological exaggeration thereof. (From kyphos (Gk): bent or bowed forward.)

L

L	The abbreviation pertaining to the vertebrae or nerves of the loin (Lumbar). E.g. L5: denotes the fifth lumbar vertebra or nerve.

Labium	A lip-like structure. Applied when the structure is paired or double. (From labium (L): a lip.)
Labrum	A rim or lip-like structure. Applied when the structure in question is unpaired. (From labrum (L): an edge or lip.)
Labyrinth	A structure consisting of interconnecting passages. (From labyrinthos (Gk): a labyrinth or maze.) Fallopius (mid-16th century) applied the term to the passages associated with hearing.
Lac	Milk. (From lac (L): milk.)
Lacerum	Lacerated, torn. (From laceratus (L): torn to pieces (From lacer (L): mangled).)
Lacertus	A term originally applied to the muscular upper arm. (From lacertus (L): the upper arm.)
Lacertus Fibrosus	The bicipital aponeurosis. (From lacertus (L): arm + fibrosus (L): fibrous. Literally, a fibrous arm.)
Lacrimal	Pertaining to the lacrimal apparatus. (From lacrima (L): a tear.)
Lacteal	Chyliferous or lymphatic vessels of the small intestine. (From lac (L): milk.) So named because of the appearance of the fluid borne.
Lacuna	A small hollow or depression. (From lacuna (L): a ditch or pond.)
Lacus	A fluid collecting space. (From lacus (L): lake.)
Lagena	The terminal portion of cochlea. (From lagena (L): a flask.)
Lamella	A thin plate-like structure. (From lamella (dim. of lamina) (L): a small plate or leaf.)
Lamina	A plate-like structure. (From lamina (L): a plate.)
Lanugo	The fine downy hair of the skin of the fetus or of the cheeks. (From lanugo (L): wool.)
Larynx	The organ just above the entrance of the trachea which produces the voice. (From laryngx (Gk): the larynx.)

Latus   (1) The side (From latus (L): the side.)
                     (2) Wide, broad. (From latus (L): broad.)

Lemniscus   A band-like structure. (From lemniskos (Gk): a band or strip.)

Lens   The transparent focusing mechanism of the eye. (From lens (L): a lentil.)

Lenticular   Lens-shaped; biconvex. (From lens (L): lentil.)

Leptomeninx   A combined term for the pia and arachnoid maters. (From leptos (Gk): thin, delicate + meningx (Gk): a membrane.) Also called the endomeninx.

Lien   The spleen. (From lien (L): the spleen.)

Lienculus   An accessory spleen. (Dim. of lien (L): the spleen.)

Ligament   A fibrous band joining two or more movable bones. (From ligamentum (L): a bandage / ligare (L): to bind.)

Ligula   A small tongue- or strap-like process. (From ligula (L): a little tongue.)

Limb   An arm or a leg. (From lim (OE): a limb.)

Limbic   Bordering. Applied to the limbic system. (From limbus (L): border, edge or hem.)

Limbus   A border. (From limbus (L): a border, edge or hem.)

Limen   A border. (From limen (L): a border or threshold.)

Linea   A line-like structure or mark. (From linea (L): a line.)

Linea Aspera   The rough line. (From linea (L): a line + aspera (L): rough.)

Lingua   The tongue. (From lingua (L): the tongue.) Also as lingualis.

Lingual   Of the tongue. (From lingua (L): the tongue.)

Lingula   A small tongue-like structure. (From lingula (dim. of lingua) (L): the tongue.)

Liquor   A bodily fluid. (From liquor (L): fluid, liquid.)

Liver   The bile-secreting organ. (From lifer (OE): the liver.)

Lobe	From lobos (Gk): a lobe.
Lobule	A small lobe. (From lobulus (L): a small lobe (from lobos (Gk): a lobe).)
Lobulus	A small lobe. (From lobulus (L): a small lobe (from lobos (Gk): a lobe).)
Locus	A place or site. (From locus (L): a place.)
Locus Caeruleus	The bluish spot on the floor of the fourth ventricle. (From locus (L): a place + caeruleus (L): sky-blue.)
Longissimus	The longest. (The superlative of longus (L): long.)
Longus	Long. (From longus (L): long.)
Lordosis	The bend of the lumbar part of the spine; more properly, a pathological exaggeration thereof. (From lordoo (Gk): I bend.)
Lumbar	The loins. (From lumbus (L): the loin.)
Lumen	The unobstructed space within a passage or vessel. (From lumen (L): light.)
Lunula	Like a small moon. (From lunula (dim. of luna) (L): a small moon.)
Lunula Unguis	The white semilunar area on the proximal part of the nail. (From lunula (L): a small moon + unguis (L): a nail or claw.)
Lung	The organ of respiration. (From lunge (OE): a lung.)
Luteum	Yellow. (From luteus (L): yellow.)
Lymph	The fluid of the lymphatic system. (From lympha (L): pure spring water.)
Lyra	A triangular structure. (From lyra (Gk): a lyre.)

M

M	The abbreviation denoting a muscle (plur. Mm.).
Macroglia	Large neuroglial cells, also called astrocytes. (From makros (Gk): large + gloia (Gk): glue.)

Macula   A spot or patch. (From macula (L): a spot.)

Malar   Pertaining to the cheek. (From mala (L): the cheek bone.) Former name for the maxilla.

Malleolus   The rounded bony eminences formed by the distal tibia and fibula on either side of the ankle joint. (From malleolus (dim. of malleus) (L): a small hammer.) The hammer alluded to is a round-headed one. This term was first used by Vesalius (16th century).

Mammary   Pertaining to the breast, inclusive of the mammary gland. (From mamma (L): the breast.)

Mammillary   Like a small breast. (From mamilla (L): a small breast or a nipple.) Also as: mammilare

Manubrium   A handle-like process. (From manubrium (L): a handle or hilt.) The term Gladiolus (L): 'a small sword' was once given to the sternum with the manubrium sterni being its handle.

Manus   The hand. (From manus (L): the hand.)

Margo   An edge, boundary or margin. (From margo (L): an edge, boundary or margin.)

Massa   A mass. (From massa (L): a mass.)

Masticatory   Pertaining to the chewing apparatus. (From masticare (L): to chew.)

Mater   A covering. (From mater (L): a mother.)

Matrix   (1) The ground substance of connective tissue;
(2) The nail bed;
(3) The uterus.
(From matrix (L): a female animal used for breeding (from mater (L): a mother).) The Roman usage indicated that the matrix enclosed something.

Maxillary   Pertaining to the upper jaw. (From maxilla (L): jaw-bone or jaw.)

Meatus   A passage or channel. (From meare (L): to go.)

Mediastinum   The central portion of the thoracic cavity between the two pleural cavities. (Derivation from mediastinus (L): a servant is questionable. Derivation via a corruption of 'per medium tensum' ('tight through the middle') has been suggested.)

Medulla   (1) The bone marrow;
(2) The innermost part of a structure. (From medulla (L): marrow.)

Membrane   A thin tissue layer or covering. (From membrana (L): a membrane.)

Meninges   The three-layered membranous covering of the brain, spinal cord and nerve roots. (From meningx (Gk): a membrane.) Plural of meninx.

Meninx   A covering of the brain or spinal cord. (From meningx (Gk): a membrane.)

Meniscus   The crescent-shaped intra-articular cartilages of the knee joint. (From meniskos (dim. of mene) (Gk): a small moon.)

Mental   Pertaining to the chin. (From mentum (L): the chin.)

Merocrine   A type of secretion in which the substance to be released is first collected as discrete units within the cell. (From meros (Gk): portion + krinein (Gk): to separate.)

Mesen-cephalon   The midbrain. (From mesos (Gk): middle + egkephalos (Gk): brain.)

Mesenchyme   The non-epithelial cells between the embryonic ectoderm and endoderm from which are derived the myocardium, the smooth muscle and the vascular and lymphoid tissues. (From mesos (Gk): middle + engchein (Gk): to pour in.)

Mesentery   A fold of peritoneum which holds viscera in place and carries vessels. (From mesenterium (L) / mesenterion (Gk): the mesentery.) This word is probably a contraction using meso (Gk): middle and intestina (L): the entrails and pertains to an intermediary structure.

Mesen-teriolum
A small mesentery. (From mesenteriolum (dim. of mesenterium) (L): a small mesentery.)

Meso- (1)
Prefix: middle. (From mesos (Gk): middle.)

-derm
The embryonic layer between endoderm and ectoderm. (From mesos (Gk) middle + derma (Gk): skin.)

-nephros
The intermediate embryonic kidney form between pronephros and metanephros. (From mesos (Gk): middle + nephros (Gk): the kidney.)

-thelium
The lining membrane derived from the embryonic mesoderm. (From mesos (Gk): middle + thele (Gk): a nipple.)

Meso- (2)
Prefix: Pertaining to a mesentery or mesentery-like structure. (From mesenterium (L) / mesenterion (Gk): the mesentery.)

-appendix
The peritoneal fold connecting the appendix and ileum. (From appendere (L): to hang on.)

-caecum
The occasional mesentery attached to the caecum. (From caecus (L): blind (blind-ended).)

-cardium
The embryonic mesentery from which the heart is suspended. (From kardia (Gk): the heart.)

-colon (transverse)
The mesentery which connects the (trans-verse) colon to the posterior abdominal wall. (From kolon (Gk): the large intestine.)

-gastrium
The embryonic mesentery, the dorsal portion of which connects the stomach to the posterior abdominal wall and, by the ventral portion, to the liver. (From gaster (Gk): the stomach.)

-metrium
That part of the broad ligament lateral to the uterus and below the mesosalpinx. (From metra (Gk): the uterus.)

-rchium	The embryonic mesentery attaching the testis to the mesonephros. (From orchis (Gk): testis.)
-rectum	The peritoneal fold of the rectum. (From rectus (L): straight.)
-salpinx	That part of the broad ligament of the uterus lying between the uterine tube, ovary and the round ligament. (From salpinx (Gk): a tube.)
-tendon	The line of reflection between the parietal and visceral layers of a tendon sheath, through which vessels and nerves pass. (From tendere (L): to stretch.)
-varium	The peritoneal connection between the ovary and the broad ligament. (From ovarium (L): ovary.)
Metaphysis	That part of the shaft of a long bone adjacent to the epiphysis (From meta (Gk): after + phyein (Gk): to grow.)
Metanephros	The final embryonic kidney form following pronephros and mesonephros which remains permanently. (From meta (Gk): after + nephros (Gk): the kidney.) See mesonephros and pronephros.
Meten-cephalon	The hindbrain. (From meta (Gk): after + kephalon (Gk): the head.)
Metopic	Pertaining to the forehead and anterior cranium. (From metopon (Gk): forehead.)
Microglia	Small neuroglial cells. (From mikros (Gk): small + gloia (Gk): glue.)
Minimus	Least. (From minimus (L): least.)
Mirabilis	Remarkable, wonderful, extraordinary. (From mirabilis (L): remarkable, wonderful, extraordinary.)
Mitral	Shaped like a bishop's mitre; bicuspid. (From mitra (L): a turban or girdle.)
Modiolus	The central bony column or axis of the cochlea. (From modiolus (L): a small Roman measure of corn.) The Roman word may also mean a trepan (a cylindrical borer with a

serrated edge) or the hub of a wheel. These may be seen to bear some resemblance to the structure in question.

Molar   A grinding or chewing tooth. (From molaris (L): relating to a mill.)

Mons   An eminence. (From mons (L): a mountain or eminence.)

Mucosa   The mucous membrane. (From mucus (L) / muxa (Gk): nasal discharge.)

Mucous   Pertaining to mucus or mucous membrane. (From mucus (L) / muxa (Gk): nasal discharge.)

Mucus   The secretion of mucous membrane. (From mucus (L) / muxa (Gk): nasal discharge.)

Multifid   Having many divisions or clefts. (From multi (L): many + fidus (L): clefts.)

Muscle   The contractile tissue responsible for movement of and within the body. (From musculus (L): a muscle; a small mouse; a kind of fish or a small ship.)

**Musculo-   A nerve of the ventral aspect of the forearm
cutaneous**   and arm supplying both muscle and skin. (From musculus (L): muscle + cutis (L): skin.)

**Myelen-   The posterior part of the hind-brain. (From
cephalon**   myelos (Gk): marrow + encephalos (Gk): the brain. Literally, the marrow brain.)

Myelin   The lipoproteinaceous sheath of myelinated nerve axons. (From myelos (Gk): marrow.) This term was first used by Virchow in the mid-19th century.

Myenteric   Pertaining to the muscle of the gut. (From mys (Gk): muscle + enteron (Gk): gut.)

Myocardium The heart muscle. (From mys (Gk): muscle + kardia (Gk): the heart.)

Myotome   A form of muscle block during embryonic development. (From mys (Gk): muscle + temnein (Gk): to cut.)

N

N	The abbreviation for nerve (plur. Nn.).
Nail	The terminal horny plate on the fingers or toes. (From naegel (OE): the nail.)
Nape	The back of the neck. (From knappe (ME): a knob. Probably referring to the external occipital protuberance.)
Nares	The nostrils, the anterior opening of the nasal cavity. (From naris (L): a nostril.)
Nasal	Pertaining to the nose. (From nasus (L): the nose.)
Nasopharynx	That portion of the pharynx continuous with the posterior nares. (A hybrid term from nasus (L): nose + pharynx (Gk): gullet.)
Nates (Natal)	The buttocks. (From nates (L): the buttocks.)
Navel	The umbilicus. (From nafele (OE): navel.)
Neopallium	That part of the cerebral cortex not concerned with the sense of smell and so being evolutionarily newer than the archipallium. (From neos (Gk): new + pallium (L): a cloak.)
Nephron	The functional unit of the kidney; the renal corpuscle and the straight and convoluted tubules. (From nephros (Gk): the kidney.)
Nephros	The kidney. (From nephros (Gk): the kidney.)
Nerve	The functional (cellular) unit of the nervous system. (From nervus (L) / neuron (Gk): a sinew.) Originally, the term was applied to any whitish, fibrous structures.
Neural	Pertaining to nerves. (From neuron (Gk): a nerve.)
Neuraxon	The central cylinder of a myelinated nerve fibre. (From neuron (Gk): a nerve + axis (L): an axis or axle.)
Neurilemma/ Neurolemma	A cell that wraps around one or more nerve axons forming the sheath of Schwann. (From neuron (Gk): nerve + lemma (Gk): a husk or wrapping.)

Neuroglia   Cells related to neurons, previously thought to be the connective tissue of the central nervous system but now believed to play an important role in neuron metabolism. (From neuron (Gk): nerve + gloia (Gk): glue.) Cell types include:
   astrocytes
   ependymal cells
   microglia
   oligodendrocytes

Neuron   A nerve. (From neuron (Gk): a sinew.) Originally, the term was applied to any whitish, fibrous structures.

Nidus   A nest-like hollow. (From nidus (L): a nest.)

Niger   Black or dark. (From niger (L): black.) Also as: nigra, nigrum.

Nipple   The teat of the breast. (Dim. of nib or neb (OE): a nose.)

Node   A knob-like structure. (From nodus (L): a knob.)

Nodulus   A small node or knob-like structure; nodule. (From nodulus (dim. of nodus) (L): a small knob.)

Norma   Anatomically, this term is used to denote the standard viewing points of the skull as a whole. (From norma (L): a rule; a square used in carpentry.)

Nose   The projection from the centre of the face bearing the nostrils. (Possibly from nosu (OE) / nasus (L): the nose.)

Nostril   The external nasal openings. (From nosthyrl (OE): nostril.)

Notochord   The dorsal supporting axis of the embryo. (From noton (Gk): back + chorde (Gk): a cord.)

Nucha   The back or nape of the neck; applied to the ligaments between the spinous processes of the cervical vertebrae. (From nookah or nuqrah (Ar): the spine or spinal cord / or possibly from nucha (LL): the spinal marrow.)

Nucleus A central delimited structure. (From nucleus (L): a nut or kernel of a nut.)

Nutricus Nutrient; nourishing. (From nutrix (L): a wet nurse.) Pertaining to the nutrient foramen.

Nymphae The labia minora. (From nympha (Gk): nymph.) The ancient Greeks used this term to denote the clitoris only. The labia minora were referred to as the pterygomata (wings). How the transition in terminology occurred is unclear.

O

Obex A small triangular fold of grey matter at the caudal end of the roof of the fourth ventricle. (From obex (L): a bolt, barrier or bar.)

Oblongata Oblong. (From oblongatus (L): oblong or elongated.)

Obturator Pertaining to the obturator foramen of the anterior pelvis. (From obturare (L): to stop up or to close.) Thus the obturator membrane is a closure

Occiput The occipital region of the skull. (From ob (L): opposite + caput (L): the head, the back of the head.)

Occlusion Closing together. (From occludere (L): to close, to shut in.) Applied to the plane where the upper and lower grinding surfaces of the teeth fit together

Ocular Pertaining to the eye. (From oculus (L): the eye.)

Oculus The eye. (From oculus (L): the eye.)

Oesophagus The tube that carries food from the pharynx to the stomach. (From oisophagos (Gk): the gullet (from oisein (Gk): to carry + phagma (Gk): food).)

Olecranon   The proximal portion of the ulna. (From olene (Gk): elbow + kranion (Gk): head.) The original term properly applies to the tip or point of the elbow now referred to as the olecranon process

Olfactory   Pertaining to the sense of smell. (From olfacere (L): to smell.)

Oligodendro-cyte   One of the types of neuroglial cells. (From oligos (Gk): few, dendron (Gk): tree + kytos (Gk): cell. Literally, cells with few branches.)

Oligodendro-glia   One of the types of neuroglial cells. (From oligos (Gk): few, dendron (Gk): tree + gloia (Gk): glue. Literally, glue (cells) with few branches.)

Oliva   An olive. (From oliva (L): an olive.)

Olivary nucleus   An ovoid protuberance of the medulla oblongata, between the pyramid and inferior cerebellar peduncle. (From oliva (L): an olive.)

Omentum   A double layer (fold) of peritoneum which may be free or may act as a connection between certain viscera and which may contain much fat. (Probably from operimentum (L): a fatty membrane or cover.) The origin of this term is unclear. An alternative etymology may be that this term is derived from omen (L): an omen, as derived from the study of animals' entrails

Omo   Pertaining to the shoulder. (From omos (Gk): the shoulder.)

Operculum   Those parts of the cerebrum which lay over the insula. (From operire (L): to cover.)

Ophthalmic   Pertaining to the eye. (From ophthalmos (Gk): the eye.)

Ora   An edge or boundary. (From ora (L): an edge or margin.)

Oral   Pertaining to the mouth. (From os (L): mouth.)

Orbicularis   A structure resembling a small circle, ring or wheel. (From orbis (L): a circle, ring or wheel.) Applied to muscle fibres encircling an opening

Orbit The eye-socket. (From orbita (L): a rut or wheel track (related to orbis (L): wheel etc.).) The Roman word 'orbita' never meant the 'eye-socket'. This term appears to have arisen due to faulty Latin usage by some Medieval anatomists

Orchis The testis. (From orchis (Gk): the testis.)

Organ Any part or structure within an organism adapted for a specific purpose. (From organum (L): an implement or instrument.)

Orifice An opening or aperture. (From os, oris (L): mouth + facere (L): to make.)

Os, Oris The mouth. (From os, oris (L): a mouth.)

Os, Ossis A bone. (From os, ossis (L): a bone.)

Os Chelonii The tortoise (shaped) bone. A term formerly applied to the axis (2nd cervical vertebra) because of its likeness to a tortoise with an outstretched neck. (From os (L): a bone + chelone (Gk): a tortoise.)

Ossicle A little bone, especially applied to the three small bones of the middle ear. (From ossiculum (dim. of os) (L): a little bone.)

Ossification The process whereby new bone is formed. (From os (L): bone + facere (L): to make.)

Ostium An opening or an orifice. (From ostium (L): a door.)

Otic Pertaining to the ear. (From ous (Gk): the ear.)

Otolith The tiny calcareous particles of the inner ear concerned with the mechanism of balance, being associated with the terminals of the vestibular nerve. (From ous (Gk): the ear + lith (Gk): a stone.)

Ovalis Oval or egg-shaped. (From ovum (L): an egg.)
Also as: ovali

Ovary	The female reproductive gland from which the egg (ovum) is expelled. (From ovarium (L): the ovary.)
Oviduct	The uterine (formerly Fallopian) tube which carries eggs from the ovary to the uterus. (From ovum (L): an egg + ductus (L): a tube or duct.)
Ovum	An egg. (From ovum (L): an egg.)
Oxyntic	Applied to the acid-producing cells of the stomach. (From oxyein (Gk): to make acid.)

P

Pachymeninx	The dura mater. (From pachys (Gk): hard or thick + meningx (Gk): a membrane.) Also called the ectomeninx.
Palate	The roof of the mouth. (From palatum (L): the palate.)
Pallidus	Pale. (From pallidus (L): pale.)
Pallium	The cerebral cortex. (From pallium (L): a cloak, mantle or coverlet.)
Palm	The flat of the hand. (From palma (L): the palm.)
Palpebra	An eyelid. (From palpebra (L): an eyelid (from palpitare (L): to move quickly (i.e. as in blinking)).)
Pancreas	A compound racemous gland on the posterior abdominal wall which produces the endocrine insulin and exocrine digestive enzymes. (From pan (Gk): all + kreas (Gk): flesh.)
Panniculus	Pertaining to the panniculus adiposus, the fat-containing superficial fascia. (From pannus (L): cloth.)
Papilla	A nipple-shaped structure. (From papilla (L): a nipple or teat.)
Para-	Prefix: Beside or near. (From para (Gk): beside or near.)
-chordal	Beside the notochord of the embryo; the cartilaginous plates that go into the formation

	of the base of the skull. (From chorde (Gk): a chord.)
-didymus	The aggregation of convoluted tubules above the head of the epididymus. (From didymos (Gk): twinned or paired. Didymos came to mean testicle.)
-flocculus	Beside the flocculus of the cerebellum. (From floccus (Gk): fleece.)
-metrium	Applied to the connective tissue partially covering the uterus. (From metra (Gk): uterus.)
-sympathe-tic	That part of the autonomic nervous system antagonistic to the sympathetic part. (From sympathetikos (Gk): sympathetic.)
-thyroid	Pertaining to the four small brownish-red endocrine glands situated immediately posterior to the thyroid gland. (From thyreos (Gk): shield-like.)
Parenchyma	The functional cellular part of an organ as opposed to its connective matrix. (From parengchyma (Gk) (from para (Gk): beside or near + engchyma (Gk): an infusion): a pouring-out into the surrounds.) This term comes from the ancient view of organ physiology.
Parietal	Pertaining to a wall-like structure. (From paries (L): a wall.) The term 'parietes' may be used to denote the thoracic or abdominal walls.
Parotid	The paired salivary glands situated between the ramus of the mandible and the external auditory meatus. (From para (Gk): beside or near + ous (Gk): the ear.)
Pars	A part. (From pars (L): a part.)
Parvus	Small. (From parvus (L): small.)
Pecten	A comb-like structure. (From pecten (L): a comb.)

Pectoral   Pertaining to the chest. (From pectus (L): the breast.)

Pedicle   The traversing structures between the vertebral body and the neural arch. (From pediculus (dim. of pes) (L): a small foot.)

Pedis   Pertaining to the foot. (From pes (L): a foot.)

Peduncle   Bands of white fibres joining different parts of the brain. (From pedunculus (a variant of pedicle) (L): a stem- or stalk-like structure.)

Pellucidus   Transparent or translucent. (From perlucere (L): to transilluminate.)
                      Also as: pellucidum.

Pelvis   A contained cavity-like structure. (From pelvis (L): a basin.)

Penicillum   Pertaining to the brush-like branches of the follicular arteries of the spleen. (From penicillum (L): a paint-brush.)

Penis   The male organ of copulation. (From penis (L): the penis or a tail.)

Pennatus   Pennate; feather-like. (From penna (L): a feather.)

Perforans   Perforating. (From perforare (L): to pierce through.)
                      Also as: perforata.

Perineum   The anatomical region bounded, anteriorly, by the pubic arch and posteriorly by the ischial tuberosities, coccyx and sacrotuberous ligaments. (From perinaion (Gk): the parts between the scrotum and the anus.)

Peri-   Prefix: around. (From peri (Gk): around.)

   -cardium   The connective tissue membrane covering the heart. (From kardia (Gk): the heart.)

   -chondrium   The fibrous connective tissue membrane covering of cartilage. (From chondros (Gk): cartilage.)

   -cranium   The fibrous connective tissue membrane covering the cranium. (From kranion (Gk): skull.)

-desmium	The connective tissue covering of a ligament. (From desmos (Gk): a band.)
-didymus	The tunica albuginea; the white fibrous covering of the testis. (From didymos (Gk): twinned or paired. Didymos came to mean testicle.)
-lymph	The fluid between the membranous labyrinth and the bony labyrinth wall in the ear. (From lympha (L): spring water.)
-metrium	The peritoneum covering the uterus. (From metra (Gk): uterus.)
-mysium	The connective tissue enclosing a fasiculus of muscle fibres. (From mys (Gk): muscle.)
-nephrium	The connective tissue membrane covering the kidney. (From nephros (Gk): kidney.)
-neurium	The connective tissue membrane covering a fasiculus of nerve fibres. (From neuron (Gk): nerve.)
-odontium	The connective tissue membrane around the root of a tooth. (From odous (Gk): tooth.)
-osteum	The fibrous connective tissue membrane covering of bone. (From osteon (Gk): bone.)
-toneum	The two-layered serous membrane covering the walls (parietal layer) and viscera (visceral layer) of the abdomen. (From teinein (Gk): to stretch.)
Peroneal	Pertaining to the fibula. (From perone (Gk): a pin.)
Peroneus	Pertaining to the fibula. (From perone (Gk): a pin.)
Pes	A foot. (From pes (L): a foot.)
Pes Anserinus	A goose's foot (From pes (L): a foot + anser (L): a goose.) Used anatomically to refer to (i) the common insertion of the sartorius, semitendinosus and gracilis muscle into the supero-medial aspect of the tibia and (ii) the plexus formed by the facial nerve as it passes through the parotid gland.

Pes Anserinus Minor   The lesser goose's foot. (From pes (L): a foot + anserinus (L): a goose + minor (L): lesser.) A term once used to describe the plexus formed by the infra-orbital nerve after its exit from the infra-orbital foramen.

Petiolus   A small stalk or pedicle. (From petiolus (L): a little foot.)

Petrosal   Pertaining to the petrous portion of the temporal bone. (From petros (Gk): a rock.)

Phallus   The indeterminate embryological structure that becomes the penis or clitoris. (From phallos (Gk): a phallus) This term is not strictly synonymous with the penis.

Pharynx   The part of the alimentary tract behind the nasal cavities and mouth and the larynx. (From pharyngx (Gk): the gullet.)

Philtrum   The vertical midline groove between the upper lip and the nose, externally. (From philtron (Gk): a love potion, or from filtrum (L): a filter.)

Phrenic   Pertaining to the diaphragm. (From phrenos (Gk): diaphragm. It has been suggested that the term phrenic is derived from phren (Gk): the mind, but the etymology is uncertain.)

Pia   Soft or tender. (From pius (L): soft or tender.)

Pilomotor   Hair moving, the action of the arrectores pilorum muscles. (From pilus (L): a hair + movere (L): to move.)

Pilus   A hair. (From pilus (L): a hair.)

Pineal   Pertaining to the pineal gland; pine cone-shaped. (From pinea (L): a pine cone.)

Pinna   The pliable, fleshy external ear. (From pinna or penna (L): a feather; especially the larger ones.) This term also became applicable to a wing.

Pituitary   The hypophysis, the gland at the base of the brain. (From pituita (L): mucus, phlegm, rheum or slime.) This term is derived from the belief of the ancient world that the brain produced a mucous substance that was discharged from the nose via the pituitary gland.

Placenta The discoid organ connecting the fetus to the uterine wall of its mother for purposes of functional interchange. (From placenta (L): a flat, round cake.)

Placode A plate-like structure; localized embryonic ectodermal thickenings contributing to cranial nerve formation. (From plax (Gk): a plate + eidos (Gk): form.)

Planum A plane-like structure. (From planum (L): a plane.)

Plastic Having the property of maintaining a shape after deformation. (From plastos (Gk): formed.)

Pleura The serous membrane covering the lungs and lining the thoracic cavity. (From pleura (Gk): a rib.)

Plexus A network-like structure. (From plexus (L): a braid or plait / plectere (L): to plait.)

Plica A fold in a structure. (From plica (L): a fold.)

Plicae Circularis The membranous folds of the duodenum, jejunum and ileum. (From plica (L): a fold + circulus (L): a circle. Literally, circular folds.) Also called valvulae conniventes.

Pollex The thumb (From pollex (L): the thumb.)

Pollex Pedis The great toe or hallux. (From pollex (L): thumb + pes (L): foot. Literally, the thumb of the foot.) This term is now rarely applied.

Polus A pole; the end of an axis. (From polus (L): a pole.)

Pomum Adami Adam's apple. The prominent anterior ridge of the thyroid cartilage. (From pomum (L): apple + adam (Ar): man.) Originally, the term was pomum viri: Man's apple (from vir (L): a man.)

Pons The part of the rhombencephalon connecting the mesencephalon and the medulla oblongata. (From pons (L): a bridge.) This structure appears to bridge the two cerebellar hemispheres.

Ponticulus   A small bridge-like structure. (Dim. of pons (L): thus a small bridge.)

Popliteal   Pertaining to the knee. (From poples (L): the ham (posterior of the thigh).) This term, therefore, is now applied to a structure lower than properly indicated by its true meaning.

Pore   A minute channel, opening or interstice. (From poros (Gk): a channel or passage.)

Porta   A point of entry through which vessels etc. may join an organ; a hilum. (From porta (L): a gate or the city gate.)

Portal   Pertaining to (i) the portal vein (ii) the hypophyseal portal system. Both are venous structures but (i) is so named in accordance with the gate of the liver (porta hepatis) into which it passes and (ii) is so named because it was described as carrying blood from the hypophysis to the hypothalamic nuclei (from portare (L): to carry.)

Portio   A part, portion or division. (From portio (L): a part.)

Precuneus   The quadrilateral area on the medial surface of the cerebral hemispheres bounded by the cingulate, subparietal and parieto-occipital sulci. (From pre (L): before, in front of + cuneus (L): a wedge.)

Prefixation   A term applied to the arrangement of a nerve plexus that is formed from spinal nerves derived from higher up the spinal cord than usual. (From pre (L): before, in front of + fixus (L): fixed, established.)

Premolar   A bicuspid tooth situated in front of (before) the molars. (From prae (L): before + molaris (L): relating to a mill.)

Prepuce   The foreskin. (From praeputium (L): a prepuce. Possibly from prae (L): before +

	posthe (Gk): the penis. Another suggestion has been from praeputare (L): to cut away (as in circumcision).)
Pretrematic	Pertaining to the cranial surface of a branchial cleft. (From trema (Gk): a hole or perforation.)
Princeps	Chief or principal. (From princeps (L): chief or principal.)
Proboscis	A trunk-like nose. (From proboskis (Gk): a trunk.) This term is used in anatomy with reference to Cyclops fetuses, in which a trunk-like nose often develops above the single central eye.
Process	A projection from a structure. (From processus (L): an advance, a going ahead.)
Proctal	Pertaining to the anus. (From proktos (Gk): the anus.)
Proctodeum	The embryonic anal depression. (From proktos (Gk): the anus + odaios (Gk): a way.)
Prominence	A protuberance or projection. (From prominere (L): to project.)
Promontory	A projection or eminence. (From promontorium (L): a promontory.)
Pronephros	The first embryonic kidney form prior to mesonephros and metanephros. (From pro (Gk): before + nephros (Gk): the kidney.)
Proprioceptor	A neural receptor conveying information about the state of the body. (From proprius (L): one's own + capere (L): to take.)
Proprius	Proper or particular. (From proprius (L): proper or particular.)
Prosector	A dissector of anatomical material. (From pro (L): before + secare (L): to cut.)
Prosencephalon	The forebrain, including the diencephalon and the telencephalon. (From pro (Gk): before + enkephalon (Gk): the brain.)
Prostate	The gland situated immediately inferior to the base of the bladder, at the commencement of the urethra, in the male. (From pro (Gk):

before + istanai (Gk): to stand.) So named because this organ 'stands before' (in front of) the bladder.

Protuberance An eminence or bulge. (From protuberantia (L): a bulge.)

Pubes The hair in the genital area. (From pubes (L): the signs of manhood.)

Pubic Pertaining to the region of the pubic bone (os pubis). (From pubes (L): adult, the signs of manhood.)

Pudendal Pertaining to the pudendal vessels and nerves. (From pudere (L): to be ashamed.) So named because the external genitalia (of which one should be ashamed, if exposed) are supplied by these structures.

Pudendum The female external genitalia. (From pudere
Muliebre (L): to be ashamed + muliebris (L): pertaining to a woman, from mulier (L): a non-virgin woman.)

Pulmonary Pertaining to the lungs. (From pulmo (L): the lung.)

Pulmones The lungs. (From pulmo (L): the lung.)

Pulp A structure's soft contents; pulp. (From pulpa (L): the soft part of an animal or fruit pulp.)

Pulposus Pulpy or soft. (From pulpa (L): fruit pulp.)

Pulvinar A structure with a cushioned appearance. (From pulvinar (L): a cushioned couch.)

Punctum A very small point, dot or orifice. (From punctum (L): a point.)

Pupil The aperture of the iris through which light passes. (From pupilla (L): the pupil of the eye.)

Putamen A structure forming a shell or encasement. (From putamen (L): a shell or husk.)

Pyloris The pyloric orifice of the stomach which opens into the duodenum. (From pyle (Gk) a gate + ouros (Gk): a keeper.)

Pyramid A pyramid-like structure. (From pyramis (Gk): a pyramid.)

Q

Quadri-geminus
Four-fold. (From quadri (L): four + geminus (L): twinned, paired or two-fold.) Combining these two terms should properly mean eight-fold but was instead used by the Romans to mean four-fold.

R

R
Abbreviation denoting a named ramus (Plur. Rr).

Racemous
A gland type having the appearance of a bunch of grapes. (From racemosus (L): bearing clusters.)

Radicle
A small root. (From radicula (dim. of radix) (L): a small root or rootlet.)

Radix
A root. (From radix (L): a root.)

Ramulus
A small branch. (From ramulus (dim. of ramus) (L): a small root.)

Ramus
A branch-like structure. (From ramus (L): a root.)

Ranine
Pertaining to the under surface of the tongue. (From rana (L): a frog.) The ranine veins on the under surface of the tongue are so called because of their similarity with the swellings occurring in the floor of a frog's mouth when croaking.

Raphe
A line marking the fusion of certain structures. (From raphe (Gk): a seam.)

Receptaculum
A receptacle. (From recipere (L): to receive.)

Receptor
That which receives. (From recipere (L): to receive.)

Recess
A niche-like space, cleft, depression or hollow space. (From recessus (L): withdrawn.)

Rectum
That part of the distal large intestine between the sigmoid colon and anal canal. (From rectus (L): straight.) The human rectum is not straight, however. The term comes from

Galen (2nd century AD) who assumed that observations made in lower animals, some of whose rectums are straight, were applicable to Man.

Rectus   Straight. (From rectus (L): straight.)

Recurrent   A structure (usually a vessel or nerve) which bends in order to return in the direction of its origin. (From re (L): back + currere (L): to run.)

Regio   Region. (From regio (L): region.)

Renal   Pertaining to the kidneys. (From ren (L): the kidney.)

Reniculus   Literally, a small kidney, but this term has also been used with reference to the lobes of the kidney. (From reniculus (dim. of ren) (L): a small kidney.)

Respiratory   Pertaining to respiration (breathing). (From respirare (L): to respire.)

Rete   A network of small blood vessels, nerves or tubules. (From rete (L): a net.)

Rete Mirabile   The glomerulus of the kidney. Where an afferent blood vessel breaks up into numerous plexiform branches thence to recombine to form a single efferent vessel. (From rete (L): a net + mirabilis (L): remarkable.)

Reticular   Having a network. (From reticulum (L): a small net.)

Reticulum   A small network. (From reticulum (L): a small net.)

Retina   The inner, light-sensitive membrane of the eye. (The derivation is uncertain but may be from rete (L): a net, although the retina has no truly net-like appearance.)

Retinaculum   A dense connective tissue condensation which holds structures, particularly tendons, in place. (From retinere (L): to retain.)

Reunient   Uniting or joining. (From reunire (L): to unite.)

Rhinen-cephalon That part of the cerebrum concerned with the sense of smell. (From rhis (Gk): nose + egkephalos (Gk): the brain.)

Rhomben-cephalon The hindbrain, consisting of the pons, medulla oblongata and the cerebellum. (From rhombos (Gk): rhomboid + egkephalos (Gk): the brain.) So named because of its shape.

Rider's Bone An occasional ossification within the adductor longus tendon. Also known as cavalry bone and exercise bone.

Rima A cleft, fissure or slit. (From rima (L): a cleft, fissure or slit.)

Rod A slim straight cylinder-like structure. (From rod (OE): a rod.)

Rostrum A beak-like process. (From rostrum (L): a bird's beak or bill.) (An alternative meaning (platform) is not used anatomically.)

Rotundum Round. (From rotundus (L): round.)

Ruber Red. (From ruber (L): red.)

Rubor Redness. (From rubor (L): redness.)

Rudimentum Rudimentary; imperfectly developed or at an early stage of development. (From rudimentum (L): first attempt.)

Rugae Folds in the mucous membrane of certain organs. (From ruga (L): a wrinkle.)

S

S Abbreviation denoting the sacral vertebra E.g. S3: denotes the third sacral vertebra.

Sabuline Sandy. (From sabulum (L): sandy.)

Sac A sac or bag-like structure. (From saccus (L): a sac or pouch.)

Saccule A small bag-like structure. (From sacculus (dim. of saccus) (L): a small bag or purse.)

Saliva The secretion from the salivary and other mucous glands of the mouth. (From saliva (L): spittle, saliva.)

Salpinx A tube. In particular the uterine tube or the auditory tube. (From salpingx (Gk): a trumpet.)

Sanguis The blood. (From sangius (L): blood.)

Saphenous Applied to two veins, an artery and a cutaneous nerve in the leg. (From al-safin (Ar): hidden or secret.) So named because the vein is readily visible for only a small part of its course and 'hidden' for the rest. A less likely derivation, because of its absence from ancient medical writings, is from the words saphis (Gk): manifest and saphenes (Gk): visible, the implication being that the vein was so named because it is sometimes varicose and, therefore, visible.

Sarcolemma The membranous sheath that covers individual muscle fibres. (From sarx (Gk): flesh + lemma (Gk): a sheath.)

Scala A staircase or ladder. (From scala (L): a staircase / scandere (L): to climb.) This term is applied to the scala tympani and scala vestibuli of the inner ear. There are, however, no definite step-like elements present in them.

Scalp The skin and subcutaneous tissue of the top of the head from which hair grows. (From scalp (ME): the scalp.)

Sciatic Pertaining to the hip region, especially the large nerve. (From ischion (Gk): the hip-joint.) This term, from the Medieval Latin sciaticus, is derived from ischiadicus (ischiadi-kos) (Gk): 'having gout of the hip'.

Sclera The outer tough fibrous layer of the eyeball. (From skleros (Gk): hard.)

Scleratome Connective tissue situated between two myotomes; mesenchyme that will form a vertebra. (From skleros (Gk): hard + tome (Gk): cutting.)

Scleromeninx The dura mater. (From skleros (Gk): hard + meningx (Gk): a membrane.)

Scriptor	A writer. (From scriptor (L): a writer.)
Scrobiculus	The small depression of the epigastric surface immediately below the xiphoid process of the sternum. (From scrobiculus (L): a small ditch.)
Scrotum	The skin pouch containing the testicles. (From scrotum (L): a skin, or something made of hide.) So named because this structure is like a bag made of skin.
Sebaceous	Containing or secreting a fatty substance. (From sebum (L): grease or tallow.)
Sebum	The secretion of the sebaceous gland. (From sebum (L): grease or tallow.)
Sella Turcica	The pituitary or hypophyseal fossa. (From sella (L): a seat + Turcicus (L): Turkish.) Literally, a Turkish saddle. So named because of its shape.
Semen	The ejaculatory fluid containing the spermatozoa. (From semen (L): seed.)
Seminiferous	Conveying semen. (From semen (L): seed + fero (L): I carry.)
Semiovalis	Partially oval- or egg-shaped. (From semi- (L): half + ovalis (L): oval.)
Septula	A small septum or partition. (From setula (dim. of septum) (L): a small dividing wall or fence.)
Septulum	A small septum. (From septulum (dim. of septum) (L): a small wall.)
Septum	A partition. (From septum (L): a diving wall or fence.)
Serous	Watery. (From serum (L): whey.)
Serratus	Serrated or with a saw-toothed edge. (From serra (L): a saw / serratus (L): serrated.)
Serum	The clear fluid of the blood from which fibrin and cells are absent. (From serum (L): whey (the watery fluid devoid of milk solids).)
Sinciput	The supero-anterior portion of the head back to the crown and including the forehead. (From sinciput (L): half of the head (from

semi-caput).) This term has also been used as an equivalent to 'bregma'.

Sinus   A hollow structure containing a cavity, depression or a dilation. (From sinus (L): a curved surface or a bay.)

Skeleton   Usually taken to refer to the bony framework of the body but may be applied in more general terms to refer to an infra-structure. (From skeletos (Gk): dried or hard.)

Skin   The external covering of the body. (From scinn (OE): the skin.)

Skull   The bony skeletal portion of the head, the cranium. (From skulle (ME): the cranium.)

Solar Plexus   A radiating network of nerves and some ganglia situated posteriorly to the stomach and supplying viscera of the abdomen. (From sol (L): the sun + plexus (L): a braid or plait.)

Soma   The body as a whole. (From soma (Gk): the body.)

Somatic   Pertaining to the body as a whole. (From soma (Gk): the body.)

Somite   An embryonic body segment. (From soma (Gk): the body.)

Spatium   A space. (From spatium (L): a space.)

Sperm   The male fertilizing cells. (From sperma (Gk): seed.)

Spermatic   Pertaining to sperm or semen. (From spermaticus (L): spermatic.)

Sphincter   A ring-like muscle which, when contracted, closes an orifice. (From sphinkter (Gk): a binder / sphingo (Gk): 'I strangle'.)

Spinatus   Spined, having spines. (From spinatus (L): spined.)

Spine   (1) The vertebral column, as a whole or by divisions,

                   (2) A bony process, especially if long and relatively narrow.

                   (From spina (L): a spine, something with a sharp point.)

Spinosus	Spinous. (From spinosus (L): having spines.)
Splanchnic	Pertaining to the viscera. (From splangchnon (Gk): entrail.)
Spleen	The large vascular lymphatic organ lying between stomach and diaphragm in the upper left of the abdominal cavity. (From splen (Gk): the spleen.)
Splenium	The posterior border of the corpus callosum. (From splenium (L): a patch, plaster or compress.) This structure is so named because it is said to resemble a rolled-up pad.
Spondylos	A vertebra. (From sphondylos (Gk): a vertebra.)
Spongiosus	Spongy or sponge-like. (From spongia (Gk): a sponge.)
Squama	A scale- or squame-like structure. (From squama (L): a fish's scale.)
Squamous	A scale- or squame-like structure. (From squama (L): a fish's scale.)
Sternebra	An unfused segment of the sternum. (Concatenation of stern(um) and (vert)ebra.)
Stoma	A mouth-like structure. (From stoma (Gk): a mouth.)
Stomach	The distended sac-like portion of the alimentary tract. (From stomachos (Gk) / stomachus (L): gullet or oesophagus.) In the ancient world, this term was applied solely to the oesophagus or sometimes to its lower end alone, being derived from stoma (Gk): a mouth + cheo (Gk): 'I pour'. 'Gaster' was the term applied to the stomach. The naming of the stomach as such appears to have occurred relatively recently.
Stomodeum/ Stomatodeum	The embryonic oral depression. (From stoma (Gk): the mouth + odaios (Gk): a way.)
Stratum	A layer or a covering. (From stratum (L): a layer, blanket or bed cover.)
Stratum Corneum	The horny, keratinized outer layer of the epidermis. (From stratum (L): a layer + corneus (L): horny.)

Stria A narrow line, streak or groove. (From stria (L): a furrow, the flute of a column.) Also as: striatum.

Stroma The connective tissue framework of a structure. (From stroma (Gk): a blanket or bed cover.)

Subiculum An underlayer or support. (From subiculum (L): a support.)

Substantia A substance. (From substantia (L): a substance.)

Sulcus A groove or furrow. (From sulcus (L): a groove or furrow.)

Supercilium The eyebrow. (From supercilium (L): the eyebrow.)

Suprarenal Above the kidneys. (From supra (L): above + renes (L): the kidneys.)

Sural Pertaining to the calf of the leg. (From sura (L): the calf of the leg.)

Suspensory Serving to suspend. (From suspendere (L): to hang up.)

**Susten-
taculum** A supporting structure. (From sustentaculum (L): a support, prop or stay.)

**Susten-
taculum Tali** The shelf-like structure on the calcaneus upon which the talus is situated. (From sustentaculum (L): a support, prop or stay + talus (L): the ankle.) Literally, the supporter of the talus (or ankle).

Sutural Bones Small bi-lamina plates of bone situated within the cranial sutures, especially the lambdoid. Also called Wormian bones. (From sutura, suere (L): to sew.)

Sympathetic The sympathetic part of the autonomic nervous system. (From syn (Gk): with + pathos (Gk): feeling.)

Synapse The functional connection between neurons. (From syn (Gk): together + haptein (Gk): to clasp.) This term was first used by Sir Charles Sherrington (1857–1952).

Syncytium An undifferentiated, multinucleated, proto-plasmic mass. (From syn (Gk): with + kytos (Gk): a cell.)

Synergist One that assists another. (From syn (Gk): together + ergon (Gk): work. Literally, to work together.)

Synovia Synovial fluid. (From syn (Gk): together + oon (Gk) / ovum (L): egg.) A term invented by Paracelsus (early 16th century). Although there is a certain likeness between synovial fluid and egg-white, Paracelsus may not have had this comparison directly in mind since he used the same term for other quite different bodily fluids.

T

T Abbreviation pertaining to the vertebrae or nerves of the thorax. E.g. T5 refers to the fifth thoracic vertebra or nerve.

Tabatière Anatomique (Fr) The anatomical snuff box.

Tabula A table (From tabula (L): a table.)

Taenia A band- or ribbon-like structure. (From taenia (L): a band or ribbon.)

Taeniae Coli The aggregated longitudinal muscle of the large intestine. (From taenia (L); band or ribbon + kolon (Gk): the large intestine. Literally, the bands of the large intestine.)

Tapetum A carpet-like structure. Particularly where nerve fibres are seen to interlace. (From tapetum (L): a carpet or tapestry.)

Tarsus A broad, flat surface. (From tarsos (Gk): the sole of the foot.)

Tectorium A covering or roof. (From tectum (L): a covering or a roof.)

Tectum A covering or roof. (From tectum (L): a covering or a roof.)

Tegmen A covering. (From tegmen (L): a covering.)

Tegmentum   That part of the midbrain dorsal to the cerebral peduncle between the cerebral aqueduct and the substantia nigra. (From tegmen (L): a covering.)

Tela   A web-like or woven tissue. (From tela (L): a woven material or a web.)

Tela Conjunctiva   Connective tissue. (From tela (L): a woven material or a web + conjungere (L): to connect.)

Tela Elastica   Elastic tissue. (From tela (L): a woven material or a web + elastikos (Gk): impulsive.)

Telencephalon   That part of the brain comprising the cerebral hemispheres and the lamina terminalis. (From telos (Gk): end + enkephalos (Gk): brain. Literally, the endbrain.)

Temple   The region of the temporal fossa. (From tempus (L): time.) The temple is so called because the effects of time (aging) are first seen in the greying of the hair in this region.

Temporal   Applied to the temple.

Tendo   A tendon. (From tendere (L): to stretch or extend.)

Tendo Achilles   The calcaneal tendon (tendo calcaneus) being the common tendon of the soleus and gastrocnemius muscles. Named after the character in Greek mythology whose only vulnerable part was the heel that had not been wetted when he was dipped into the river Styx by his mother when a child.
(Anatomically, this term was first used in 1693 by Verheyen (Professor of Anatomy at Louvain) who dissected his own amputated leg.)

Tendo Calcaneus   See Tendo Achilles.

Tendon   A fibrous cord or band connecting a muscle to its attachment. (From tendere (L): to stretch or extend.)

Tenia	Alternative spelling of taenia.
Tentorium	A tent-like covering. (From tentorium (L): a tent.)
Tenuis	Thin, delicate. (From tenuis (L): thin, fine, delicate.)
Teres	Round or rounded off, cylindrical. (From terer (L): to rub.) The teres muscles were so named by Cowper (17th/18th century). These muscles are not, however, cylindrical. Similarly the ligamentum teres capitis femoris is not cylindrical.
Terminalis **(Terminale)**	Pertaining to an end, boundary, or limit. (From terminus (L): end.)
Testicle/ **Testis**	The male gonad. (From testiculus (L): the testis from, in turn, testis (L): a witness.) Under Roman law, no man was allowed to bear witness unless he possessed both testes.
Thalamus	The ganglionic mass in the lateral wall of the third ventricle. (From thalamus (Gk): a (bed) room or chamber.) Originally, this term was applied to certain structures in which the chambers were hollow.
Theca	A capsule or covering structure. (From theca (L) / theka (Gk): a capsule, sheath or envelope.)
Thenar	The fleshy eminence at the base of the thumb. (From thenar (Gk): the hand, the palm.) This term came to its present meaning with Rufus Ephesius (1st/2nd century AD).
Thorax	The chest. (From thorax (Gk): an item of armour covering chest and abdomen.) It was only following Plato's limitation to mean the chest region alone and the adoption of this meaning in the writings of Galen (late 1st century AD) that 'thorax' came, anatomically, to mean the chest rather than an item of armour.
Thymus	A lymphoid gland found in the lower neck and anterior portion of the superior mediastinum

in prepubescent children. (From thymos (Gk): sweetbread.) An alternative translation may be from thymos (Gk): vital force or soul, and may be relevant to the nearness of the gland to the heart.

Tissue

The arrangement of cells and associated matrix into the fundamental structural material of the body. (From tissu (Fr): woven.)

Tongue

The mobile protrusible organ of the floor of the mouth. (From tunge (OE): the tongue.)

Tonsil

An aggregation of lymphoid tissue near the base of the tongue. (From tonsilla (L): a sharpened stake.) (Why this term is applied as it is anatomically is unclear.)

Tonsilla Tubarii

The lymphoid tissue at the opening of the auditory tube. (From tonsilla (L): (see tonsil) + tuba (L): a trumpet, tube or pipe.)

Torus

A swelling or an eminence. (From torus (L): a swelling or protuberance.)

Trabecula

Structures resembling a small beam. (From trabecula (dim. of trabs) (L): a small wooden beam.)

Trabs

A beam-like structure. (From trabs (L): a wooden beam or the rib of a ship.) (Also trapes (Gk): a large wooden beam.)

Trachea

The windpipe, extending from the larynx to its point of bifurcation. (From tracheia (Gk): rough.) It was the belief in the ancient world that the arteries carried air. Thus, to distinguish between the arteries and the trachea, the latter was so named because of unevenness of its lumen caused by its cartilaginous rings.

Tract

That which connects various structures, e.g. a pathway within the nervous system. (From tractus (L): a length of wool drawn out in the process of spinning (trahere (L): to draw.))

Tractus

A tract. (From tractus (L): a length of wool drawn out in the process of spinning (From trahere (L): to draw.)

Tragus   The posteriorly projecting flap of the auricle which covers the external auditory meatus. (From tragus (L); tragos (Gk): a goat.) It is believed that this term derives from the resemblance between the tufts of hair on this structure and a goat's beard.

Transversus   Transverse, crossing-over. (From trans (L): across + versus (L): to turn.)

Triceps Surae   A name applied to the combined gastrocnemius and soleus muscles since they share a common tendon of insertion. (From tres (L): three, -ceps (L): heads + sura (L): the calf of the leg. Literally, the three-headed (muscle) of the calf.)

Tricuspid   Having three cusps or pointed projections. (From tres (L): three + cuspis (L): a point.)

Trigone   A triangular structure. (From trigonum (L): a triangle.)

Triticeous   Resembling a grain of wheat. (From triticum (L): a grain of wheat.)

Trochanter   The large bony eminences of the proximal femur. (From trochanter (Gk): a runner.) This term appears to be derived from an understanding of the functional significance of these protuberances.

Trochlea   Any supposedly pulley-like structure. (From trochilea (Gk): a pulley; trochlea (L): a pulley-block mechanism for the raising of objects.) The structures that this term is applied to and its proper functional meaning do not necessarily fit. The trochlea of the superior oblique muscle (of the eye) was so named by Aurentius in 1587.

Trochlear   Relating to a 'pulley-like' structure. (From trochilea (Gk): a pulley; trochlea (L): a pulley-block mechanism for the raising of objects.)

Truncus   (1) The body excluding the head, arms and legs.

(2) The main stem of a vessel or nerve. (From truncus (L): the trunk of a tree.)

Tuba   A tubular structure. (From tuba (L): a trumpet, tube or pipe.)

Tube   A tubular structure. (From tuba (L): a trumpet, tube or pipe.)

Tuber   An eminence or a swelling. (From tuber (L): a knob, swelling or eminence.)

Tubercle   A small eminence or swelling. (From tuberculum (L): a small hump.)

Tuberosity A small rounded eminence on a bone. (Dim. of tuber (L): a knob, swelling or eminence.)

Tubule   A small tubular structure. (From tubulus (L): a small canal.)

Tubulus   A small tubular structure. (From tubulus (L): a small canal.)

Tunica   A tissue covering; the 'coat' of a structure. (From tunica (L): the principal Roman undergarment.)

Tympanum   The tympanic cavity of the middle ear. (From tympanum (Gk): a tamborine-like drum.) Fallopius (16th century) first called the middle ear and the membrane between it and the external ear: tympanum.

U

Ultimus   Final or ultimate. (From ultimus (L): final or ultimate.)

Umbilicus   The navel. (From umbilicus (L): the navel.)

Umbo   An elevation or boss. (From umbo (L): an elevation or boss, as at the centre of a shield.)

Uncus   A hook-like structure. (From uncus (L): a hook.) Also as: uncinatus.

Ungual   Having digits with nails or claws. (From unguis (L): a nail, claw or talon.)

Unguiculus A small nail. (From unguiculus (dim. of unguis) (L): a small nail.)

Unguis
A nail of finger or toe. (From unguis (L): a nail, claw or talon.)

Unilocularis
Having one compartment or division. (From uni- (L): one + loculus (dim. of locus) (L): a small place.)

Urachus
The canal connecting the bladder and the umbilicus in the fetus. (From ouron (Gk): urine + echein (Gk): to hold.)

Ureter
The duct conveying urine from the kidney to the bladder. (From oureter (Gk): the ureter.)

Urethra
The duct conveying urine out from the bladder and additionally in the male semen. (Ourethra was a term invented by Hippocrates (5th century BC) being derived from ouron (Gk): urine.) Formerly, the urethra was thought to be bifid: one duct passing urine, the other semen.

Urine
The fluid excretion of the kidneys. (From urina (L): urine.)

Uriniferous
Carrying urine. (From urina (L): urine + fero (L): I carry.)

Uterus
The womb. (From uterus (L): the womb.) In ancient times, it appears that this term may only have been applied to the pregnant state of the organ. A Roman 'uter' was a large inflatable animal skin bag.

Utricle
The sac of the membranous labyrinth of the internal ear, lying in the vestibule. (From utriculus (dim. of uterus) (L): a small womb.)

Uvea
The pigmented, vascular layer of the eyeball comprising the choroid, ciliary body and the iris. (From uva (L): a grape.) This term derives from the resemblance of this area to a grape that has had the stalk removed, thereby leaving a hole corresponding to the pupil.

Uvula
(1) A lobe of the cerebellum.
(2) The pendulous process at the back of the soft palate. This term dates from the late 17th century, although Celsus (early 1st

century AD) uses the term 'uva'. 'Uva' was usually used to denote the inflamed condition of the uvula.

(3) The median elevation of the bladder adjacent to the internal urethral orifice. (From uvula (dim. of uva) (L): a small grape.)

V

V	(Abbreviation) Vein (plur. Vv).
Vagina	The passage leading from the uterus to the external genital opening; the birth canal. (From vagina (L): a sheath or scabbard.)
Valgus	Where the calves of the legs are outwardly bowed. (From valgus (L): bow-legged.) Opposite of varus.
Vallecula	A depression, groove or fossa. (Dim. of vallis (L): a valley.)
Vallum	A wall or rampart. (From vallum (L): a wall or rampart.)
Vallum Unguis	The fold of skin which covers the sides and proximal ends of the nail. (From vallum (L): a wall + unguis (L): a nail or claw.)
Valve	A structure preventing contraflow of fluid. (From valva (L): a fold.)
Valvula	A valve. (From valva (L): a fold.)
Valvulae Conniventes	The membranous folds of the duodenum, jejunum and ileum. (From valva (L): a fold + connivere (L): to wink. Literally, converging folds.) Also called plicae circularis.
Varus	Where the calves of the legs are inwardly bowed. (From varus (L): knock-kneed.) Opposite of valgus.
Vas	A small vessel, duct, canal or blind tube. (From vas (L): a vessel or dish.)
Vasculosa	Vascular, containing vessels. (From vas (L): a vessel.)
Vaso-	Prefix: Pertaining to vessels. (From vas (L): a vessel.)

-constriction	A reduction in a vessel's lumen. (From constringere (L): to draw tight.)
-dilation	An increase in a vessel's lumen. (From dilatus (L): separated.)
-motor	Applied to nerves driving the action of vasoconstriction and vasodilation. (From movere (L): to move.)
Vein	A blood vessel transporting blood towards the heart. (From vena (L): a blood vessel or vein.)
Velamentum	A membranous covering. (From velare (L): to cover.)
Veli	Of a veil. (From velum (L): a veil, sail or covering.)
Velum	A membranous structure similar to a veil. (From velum (L): a veil, sail or covering.)
Vena	A vein. (From vena (L): a vein.)
Venae Comitantes	The veins that accompany an artery. (From venae (L): veins + comitari (L): to accompany.)
Venter	A belly. (From venter (L): a belly or paunch.)
Ventricle	A chamber or cavity of the heart or within the brain. (From ventriculus (dim. of venter) (L): a small belly.)
Venule	A small vein. (From venula (dim. of vena) (L): a small vein.)
Vermis	The narrow median structure connecting the two cerebellar hemispheres. (From vermis (L): a worm.)
Vernix	Varnish. (From vernix (Mod L): varnish. Probably from bemike (Med Gk): varnish or possibly from vernatio (L): shedding of skin.)
Vernix Caseosa	The fatty substance that covers a fetus. (From vernix (Mod L): varnish + caseosa (L): cheesy.)
Vertex	A whorl-like or spiral arrangement. Also used to mean a summit or highest point. (From vertex (L): a whorl.)

Verumon-
tanum
The protrusion of the floor of the prostatic urethra. (From veru (L): a spike or ridge + montanus (L): mountainous.)

Vesica
The bladder. (From vesica (L): a bladder.)

Vesicle
A small bladder-like structure. (From vesicula (dim. of vesica) (L): a small bladder.)

Vestibule
A cavity or passage preceding entry into another space. (From vestibulum (L): a forecourt or entrance.)

Vestige
A degenerate or imperfectly formed structure which may have been complete at a previous evolutionary stage. (From vestigium (L): a trace or token.)

Vibrissae
A nostril hair. (From vibrissa (L): a nostril hair / vibrare (L): to vibrate.)

Villus
Minute vascular processes to be found in the intestine, placenta etc. (From villus (L): shaggy hair.)

Vinculum
A cord- or band-like structure. (From vinculum (L): a binding, cord or chain.)

Viscus
The internal organs, especially of the abdomen and thorax. (From viscera (L): the bowels.)

Visceral
Relating to a vicus or the viscera. (From viscera (L): the bowels.)

Vitellus
The egg yolk. (From vitellus (L): a little calf.) The term was transferred by Celsus (early 1st century AD).

Vitreous
Transparent. (From vitreus (L): glassy.)

Volar
Pertaining to the palm of the hand or the sole of the foot. (From vola (L): palm of the hand.)

Vortex
A whirl-like or spiral arrangement. (From vortex (L): a whirlpool.)

Vox
The voice. (From vox (L): the voice.)

Vulva
The external female genitalia. (From volva (L): a covering or wrapper.) This term was originally applied to the uterus but was later applied to the external female genitalia only

because, according to Isidorus (6th/7th century AD), of the resemblance to the flaps of a folding door (valvae).

Z

Zona   A zone. (From zone (Gk): a belt, girdle.)

Zonula   A small zone. (From zonule (dim. of zone) (Gk): a small girdle.)

Zygapophysis A vertebral articular process. (From zygotos (Gk): yoked + apo (Gk): from + physis (Gk): growth.

Zygote   The fertilized diploid cell resulting from the union of a sperm and an ovum. (From zygotos (Gk): yoked.)

Eponyms

Many anatomical structures have the names of previous workers applied to them. For example, the atrio-ventricular bundle is commonly called the bundle of His, after Wilhelm His (1863–1934). There follows a short and by no means exhaustive list of some of the more commonly used proper names that have been lent to particular anatomical structures and a brief biographical note on the person after whom the structures were named.

ADDISON, Christopher (1869–1951) (Lord Addison of Stallingbrough) : British anatomist and politician. First Minister of Health (1919).
  Addison's Transpyloric Plane — the upper transverse abdominal line.

ALCOCK, Benjamin (1801–?) : Irish anatomist.
  Alcock's Canal — the canal of the internal pudendal vessels in the ischio-rectal fossa.

AUERBACH, Leopold (1828–1897) : Physician and neuropathologist in Breslau.
  Auerbach's Plexus — the myenteric plexus.

BALL, Charles Bent (Sir) (1851–1916) : Irish surgeon.
  Ball's Valves — the rectal valves.

BARTHOLIN, Caspar Secundus (1655–1738) : Professor of Medicine, Anatomy and Physics in Copenhagen.
  Bartholin's Glands — greater vestibular glands.

BELLINI, Lorenzo (Laurentio) (1643–1704) : Physician and Professor of Philosophy and Anatomy at Pisa.
  Bellini's Tubules — of the kidney.
  Bellini's Ducts — the orifices of the tubules.

BETZ, Vladimir Aleksandrovich (1834–1894) : Professor of Anatomy at Kiev.
  Betz Cells — the large pyramidal cells of the motor cortex.

BIGELOW, Henry Jacob (1818–1890) : Professor of Surgery at Harvard University Medical School.

Bigelow's Ligament — The ilio-femoral ligament.

BIRMINGHAM, Ambrose (1864–1905) : Professor of Anatomy at the Catholic University of Ireland.

Birmingham's Stomach Bed — the surface of the organs against which the posterior surface of the stomach lies.

BOWMAN, William (Sir) (1816–1892) : Ophthalmic surgeon and Professor of Anatomy and Physiology at King's College, London.

Bowman's Capsule — the surrounding capsule of the glomerulus of the kidney.

BROCA, Pierre Paul (1824–1880) : French surgeon, anatomist and anthropologist.

Broca's Gyrus (or Convolution) — the 'speech area' in the cerebral cortex.

BRODEL, Max (1870-1941) : Associate Professor and Director of the Institute of Art as Applied to Medicine, Baltimore.

Brodel's Bloodless Line (on the kidney) — the line dividing the areas supplied by the anterior and posterior branches of the renal artery.

BRUNNER, Johann Konrad (1653–1727) : Swiss anatomist. Professor of Anatomy in Heidelberg and later Strasbourg.

Brunner's Glands — digestive glands of the duodenum.

CAMPER, Peter (1722–1789) : Dutch physician, anatomist, palaeontologist, anthropologist and artist. Holder of numerous chairs.

Camper's Fascia — the superficial layer of the superficial fascia of the abdomen.

CLELAND, John (1835–1925) : Scottish anatomist. Professor of Anatomy at Glasgow.

Cleland's Cutaneous Ligaments — of the digits.

COLLES, Abraham (1773–1843) : Irish anatomist and surgeon. Professor of Anatomy and Surgery at Dublin.

Colles Fascia — the peroneal fascia and the deep layer of the superficial fascia of the abdomen.
Colles Ligament — the reflected part of the inguinal ligament.

COOPER, Astley Paston (Sir) (1768–1841) : Eminent English surgeon and anatomist. Professor at Surgeon's Hall and President of the Royal College of Surgeons.
Cooper's Ligament — (a) the upper part of the pectineal fascia.
(b) the transverse band of the ulnar collateral ligament.

CORTI, Alfonso (Marquis) (1822–1888) : Italian histologist.
Organ of Corti — the auditory hair cells in the cochlea.

COWPER, William (1666–1709) : Surgeon and anatomist in London.
Cowper's Glands — the bulbo-urethral glands.

CUVIER, Georges Leopold Chrétien Frédéric Dagobert (Baron) (1769–1832) : Zoologist, Palaeontologist and Naturalist and Professor of Natural History in Paris
Cuvier's Duct — The terminal junction of the cardinal veins in the embryo.
Cuvier's Canal — The sinus venosus.

DARWIN, Charles Robert (1809–1882) : English naturalist. Writer of 'The Origin of Species' etc.
Darwin's Tubercle — occasionally present on the helix of the pinna. (See Woolner.)

DOUGLAS, James (1675–1742) : Scottish anatomist and 'man-midwife' in London.
Pouch (or cul-de-sac) of Douglas — the recto-uterine peritoneal pouch.
Line of Douglas — the linear semicircularis of the deep aspect of the rectus abdominus sheath.
Fold of Douglas — the recto-uterine folds.

EDINGER, Ludwig (1855–1918) : Anatomist and neurologist in Frankfurt-am-Main.
Edinger's Nucleus — the Edinger-Westphal (parasympathetic) nucleus of the third cranial nerve.

ERB, W.H. (1840-1921) : Physician and neurologist in Heidelberg.

Erb's Point — the point of emergence of the upper brachial plexus on the posterior border of the sterno-cleido-mastoid muscle.

EUSTACHI (Eustachio, Eustachius), Bartolomeo (1500/10/13/ 24–1574) : Professor of Anatomy at Rome and physician to the Pope.

Eustachian Tube — the auditory tube.

FALLOPPIO, Gabriele (FALLOPPIUS or FALLOPIA, Gabriel) 1523–1562/63) : Professor of anatomy and surgery at Padua. Also Professor of Botany.

Fallopian Tubes — the uterine tubes.

GALEN (GALENUS, Claudius or Clarissimus) (130–200 AD) : Practised medicine in Rome. Physician to Marcus Aurelius in Venice. Founder of the 'Galenic System' of medicine.

Great Vein of Galen — the great cerebral vein.

GASSER, Johann Laurentius (1723–1765) : Professor of Anatomy at Vienna.

Gasserian Ganglion — the trigeminal ganglion.

GLISSON, Francis (1597–1677) : English physician and anatomist. Professor of Physic at Cambridge. One of the founders of the Royal Society. President of the Royal College of Physicians.

Glisson's Capsule — the fibrous capsule of the liver.

GOLGI, Camillo (1844–1926) : Professor of Histology and Anatomy at Pavia and at Siena.

Golgi Corpuscles — sensory encapsulated nerve-endings.

GRAAF, Regnier de (1641–1673) : Anatomist and later physician at Delft.

Graafian Follicles — an ovarian follicle containing an ovum and follicular fluid.

HARTMANN, Robert (1831–1893) : Anatomist and anthropologist. Professor of Anatomy at Berlin.

Hartmann's Pouch — a dilatation at the neck of the gall-bladder.

HAVERS, Clopton (1650/55/57–1702) : Physician in London.
  Haversian Canals — the vascular canals within bone.

HENLE, Friedrich Gustav Jakob (1809–1885) : Prosector of Anatomy at Berlin. Professor of Anatomy at Zurich and Göttingen.
  Loop of Henle — the U-shaped portion of the uriniferous tubules of the kidney.

HIGHMORE, Nathaniel (1613–1685) : English physician and botanist.
  Antrum of Highmore — the maxillary sinus.

HILTON, John (1805–1878) : Surgeon at Guy's Hospital, Hunterian Professor of Anatomy at the Royal College of Surgeons of England, being one of its original fellows and later President.
  Hilton's Law — The motor nerve to a muscle also supplies the joint which that muscle moves and its overlying skin.
  Hilton's White Line — the interval between the internal and external anal sphincters.

HIS, Wilhelm (Junior) (1863–1934) : Professor of Anatomy at Leipzig, Basle, Gottingen and Berlin.
  Bundle of His — the atrio-ventricular bundle.

HOWSHIP, John (1781–1841) : Surgeon in London.
  Howship's Lacunae (Foveolae or Pits) — resorptive recesses in bone occupied by osteoclasts.

HUNTER, John (1728–1793) : Scottish surgeon and anatomist. Spent his entire working life at St George's Hospital, London. Founder of the Hunterian Museum of the Royal College of Surgeons of England.
  Hunter's Canal — the adductor (or sub-sartorial) canal of the thigh.

KERCKRING, Theodore (1640–1693) : Physician and anatomist in Amsterdam.
  Kerckring's Valvules (or Folds) — the circular mucous folds of the small intestine.

KILLIAN, Gustav (1860–1921) : Director of the Rhinolaryngo-logical Clinic at Freiburg.
The Dehiscence of Killian — the lowest of thyropharyngeus' fibres immediately superior to cricopharyngeus.

KUPFFER, Karl Wilhelm von (1829–1902) : Anatomist and embryologist. Professor of Anatomy at Kiel, Königsberg and Munich.
Kupffer's Cells — the phagocytic cells of the liver sinusoids.

LANGERHANS, Paul (1847–1888) : Physician and anatomist. Professor of Pathological Anatomy at Freiburg.
Islets of Langerhans — insulin-producing, endocrine cell accumulations in the pancreas.

LEYDIG, Franz von (1821–1908) : Professor of Histology at Würzburg, Tübingen and Bonn. Founder of Comparative Histology.
Leydig's Cells — the interstitial, testosterone-producing cells of the testes.

LIEBERKÜHN, Johann Nathaniel (1711–1756) : Physician and anatomist in Berlin.
Crypts of Lieberkühn (Lieberkühn's Crypts, Glands or Follicles) — the straight, tubular intestinal glands.

LITTLE, James Laurence (1836–1885) : American surgeon. Professor of Surgery at the University of Vermont.
Little's Area — of the nasal septum.

LOUIS, Pierre Charles Alexandre (1787–1872) : Physician in Paris.
Angle of Louis — the sternal angle.

LUSCHKA, Hubert (1820–1875) : Professor of Anatomy at Tübingen.
Foramen of Luschka — the lateral apertures of the fourth ventricle.

McBURNEY, Charles (1845–1913) : Surgeon in New York, especially known for surgery of the abdomen and appendix.
McBurney's Point — the surface marking of the appendix.

MAGENDIE, Francois (1783–1855) : Professor of Pathology and Physiology at the College of France and physician at Hotel Dieu.

Foramen of Magendie — the median aperture of the fourth ventricle.

MALPIGHI, Marcello (1628–1694) : Professor of Medicine at Bologna.

Malpighian Corpuscles — the splenic lymph corpuscles.

MECKEL, Johann Friedrich (1714–1774) : Professor of Anatomy, Botany and Gynaecology at Berlin.

Meckel's Ganglion — the sphenopalatine ganglion.

Meckel's Cave (or Cavity) — the dural space in which the trigeminal ganglion is situated.

MECKEL, Johann Friedrich (1781–1833) (Grandson of the former) : Professor of Anatomy and Surgery at Halle.

Meckel's Cartilage — the cartilage of the first branchial arch.

Meckel's diverticulum — the diverticulum of the ileum representing the sometimes persistent vitello-intestinal duct.

MEISSNER, Georg (1829–1903) : Professor of Anatomy and Physiology at Basle; of Zoology and Physiology at Freiburg and of Physiology at Gottingen.

Meissner's Corpuscles — tactile nerve endings found in the corium, conjunctivae, lips, etc.

MERKEL, Friedrich Sigmund (1845–1919) : Professor of Anatomy at Rostock, Königsberg and Gottingen.

Merkel's Corpuscles — tactile nerve endings in the deep layer of the epidermis.

MONRO, Alexander (Secundus) (1733–1817) : Professor of Anatomy at Edinburgh.

Foramen of Monro — the interventicular foramen connecting the lateral and third ventricles.

MONTGOMERY, William Fetherston (1797–1859) : Obstetrician in Dublin.

Montgomery's Tubercles (or Glands) — the areolar glands of the breast.

MORRIS, Henry (Sir) (1844–1926) : Surgeon in London. President of the Royal College of Surgeons.
Morris's (Kidney) Box — the surface marking of the kidney in the loin.

MÜLLER, Johannes Peter (1801–1858) : Professor of Anatomy at Bonn and of Anatomy and Physiology at Berlin.
Müllerian Canal (or Duct) — the paramesonephric ducts of the embryo. (The primordial female genital duct.)

NISSL, Franz (1860–1919) : Neurologist in Frankfurt, Heidelburg and Munich.
Nissl Bodies (or Granules) - basophilic, RNA rich granules of the cytoplasm of nerve cells.

ODDI, Ruggero (1864–1913) : Italian physiologist and physician.
Sphincter of Oddi — the sphincter of the terminal common bile duct.

OLLIER, Louis Xavier Edouard Leopold (1825/30–1900) : Surgeon in Paris and Lyon. Especially known for the surgery of bones and joints.
Ollier's Layer — the inner or osteogenic layer of the periosteum.

PACINI, Filippo (1812–1883) : Anatomist and histologist. Professor of Anatomy and Physiology at Pisa and Florence.
Pacini Corpuscles (or Bodies) — sensory end organs.

PANETH, Joseph (1857–1890) : Physiologist with chairs at Breslau and Vienna.
Paneth Cells — eosinophilic granule-containing cells of the mucosa of the small intestine at the base of the crypts of Lieberkühn.

PEYER, Johann Conrad (1653–1712) : Professor of Logic, Rhetoric and Medicine at Schaffhausen.
Peyer's Patches — aggregations of lymphoid tissue in the small intestine.

POUPART, Francois (1616–1708) : Surgeon and naturalist in Reims. Surgeon at Hotel Dieu in Paris.
Poupart's Ligament — the inguinal ligament.

PURKINJE, Johannes Evangelista (1787–1869) : Physiologist and anatomist. Professor of Physiology at Breslau and Prague.
Purkinje Fibres — the large specialized muscles forming the terminal conducting system of the heart.
Purkinje Cells — multidendritic nerve cells of part of the molecular layer of the cerebral cortex.

RANVIER, Louis Antoine (1835–1922) : Physician and histologist. Professor at the College of France in Paris.
Nodes of Ranvier — the constrictions interrupting the medullary sheaths of myelinated nerve cell axons.

RATHKE, Martin Heinrich (1793–1860) : Physiologist and pathologist. Professor of Zoology and Anatomy at Königsberg.
Rathke's Pouch — the ectodermal diverticulum which develops in the stomatodeum roof to form the anterior lobe of the pituitary gland.

REICHERT, Karl Bogislaus (1811–1883) : Professor of Anatomy at Dorpat and at Berlin.
Reichert's Cartilage — The cartilage of the second branchial arch.

REID, Robert William (1851–1939) : Demonstrator of Anatomy in London. Professor of Anatomy at Aberdeen.
Reid's Base Line — a craniometric base line of the skull, originally used for locating areas of the brain beneath.

REIL, Johann Christian (1759–1813) : Physician and anatomist. Professor of Medicine at Halle and Berlin.
Island of Reil — the insula of the cerebral cortex.

REISSNER, Ernst (1824–1878) : Professor of Anatomy at Dorpat and Breslau.
Reissner's Membrane — the vestibular membrane separating the scala vestibuli from the duct of the cochlea.

ROLANDO, Luigi (1773–1831) : Anatomist. (First) Professor of Practical Medicine at Sasseri (Sardinia) and Professor of Anatomy at Turin.
Fissure of Rolando — the central sulcus between frontal and parietal lobes of the cerebrum.

RUFFINI, Angelo (1874–1929) : Professor of Histology at Bologna.
  Ruffini's Corpuscles — tactile sensory nerve endings of the corium found especially in the digital pulp.

SANTORINI, Giandomenico (Giovanni Domenico) (1681–1737) : Anatomist and Physician. Professor of Anatomy and Medicine at Venice.
  The Duct of Santorini — the accessory pancreatic duct.

SAVAGE, Henry (1810–1900) : Gynaecologist. Lecturer in Anatomy in London.
  Savage's Peroneal Body — the peroneal body.

SCARPA, Antonio (1746/47/52–1832) : Professor of Anatomy at Pavia.
  Scarpa's Triangle — the femoral triangle.
  Scarpa's Fascia — the lower superficial abdominal fascia which fuses with the deep fascia to form the inferior boundary of the superficial peroneal pouch and the inguinal ligament.

SCHLEMM, Friedrich (1795–1858) : Professor of Anatomy at Berlin.
  Canal of Schlemm — the venous sinus of the sclera at its junction with the cornea.

SCHWANN, Theodor (1810–1882) : Professor of Anatomy at Louvain and of Comparative Anatomy and Physiology at Liège. Originator of the neuronal theory of the nervous system.
  Schwann's Sheath — the neurilemma.

SERTOLI, Enrico (1842–1910) : Professor of Experimental Physiology at Milan.
  Sertoli Cells — the cells supporting the testicular epithelium.

SHARPEY, William (1802–1880) : English anatomist. Professor of Anatomy at Edinburgh and University College, London.
  Sharpey's Fibres — connective tissues fibres running between periosteum and bone.

STENSON (STENO), Niels (1638–1686) : Professor of Anatomy at Copenhagen. Later Bishop of Titiopolis (in Greece).

Also known, from his geological work, as 'The Father of Geology'.

Stenson's Duct — the parotid duct.

SYLVIUS, Francois de la Boe (1614–1672) : German anatomist. Professor of Practical Medicine at Leyden.

Aqueduct of Sylvius — the 'aqueduct of the midbrain' between the third and fourth ventricles.

Fissure of Sylvius — the lateral cerebral fissure.

SYMINGTON, Johnson (1851–1924) : English anatomist. Professor of Anatomy at Belfast.

Symington's Anococcygeal Body — the anococcygeal body.

THEBESIUS, Adam Christian (1686–1732) : Anatomist and pathologist at Leyden.

Veins of Thebesius — the smallest veins draining the heart directly into its inner cavities — mainly the atria.

THOMSON, Allen (1809–1884) : Professor of Anatomy at Aberdeen, Edinburgh and Glasgow.

Thomson's Fascia — the iliopectineal fascia.

TREVES, Frederick (Sir) (1853–1923) : Surgeon in London, especially known for his pioneering abdominal surgery. Performed an appendectomy on Edward VII and befriended Joseph (a.k.a. John) Merrick, 'The Elephant Man'.

Treves' Bloodless Fold — of the appendix — the iliocaecal fold.

TULP, Nicolas (TULPIUS) (1593–1674) : Prosector of Anatomy at Amsterdam. Figured in Rembrandt's 'Anatomy Lesson of Dr Tulp'.

Tulp's Valve — the iliocaecal valve.

VAROLIUS (VAROLIO) Constantio (1543–1575) : Professor of Anatomy at Bologna and Rome. Physician to Pope Gregory XIII.

Pons Varoli (Varolius' Pons, or Pons of Varolius) — the pons.

VATER, Abraham (1684–1751) : Professor of Anatomy and Botany and later of Pathology and Therapeutics at Wittenburg.

Ampulla of Vater — the ampulla of the hepatopancreatic duct.

VESALIUS, Andreas (1513/4–1564) : Professor of Anatomy at Padua, Bologna and Pisa. Famous for his work illustrated by wood-cuts.

Vesalius Bone — the tuberosity of the fifth metatarsal when occurring as a separate sesamoid bone.

VOLKMANN, Alfred Wilhelm (1800–1877) : Professor of Physiology and Anatomy at Dorpat and Halle.

Volkmann's Canals — canals within bone, perpendicular to the surface, carrying blood vessels from the periosteum to the Haversian canals.

WERNICKE, Karl (1848–1905) : Professor of Neurology and Psychiatry at Berlin, Breslau and Halle.

Wernicke's Centre — the sensory speech centre of the posterior third of the superior temporal gyrus.

WESTPHAL, Karl Friedrich Otto (1833–1890) : Neurologist and psychiatrist. Director of the Brain Institute. Professor of Psychiatry at Berlin.

Westphal's Nucleus — the Edinger-Westphal (parasympathetic) nucleus of the third cranial nerve.

WHARTON, Thomas (1610/14/16–1673) : Physician in London.

Wharton's Duct — the duct of the submandibular gland.

Wharton's Jelly — the gelatinous connective tissue of the umbilical cord.

WILLIS, Thomas (1621/22–1675) : Physician. Professor of Natural Philosophy at Oxford. One of the founders of the Royal Society.

Circle of Willis — the arterial ring at the base of the brain.

WINSLOW, Jacob Benignus (1669–1760) : Danish anatomist. Professor of Anatomy (from age 74 years) and of Physic and Surgery.

Foramen of Winslow — the epiploic foramen.

WIRSUNG, Johann Georg (?–1643) : Professor of Anatomy at Padua.

Wirsung's Duct — The pancreatic duct.

WOLFF, Julius (1836–1902) : Orthopaedic surgeon and Professor of Orthopaedic Surgery at Berlin.
   Wolff's Law — that internal changes accompany functional changes in bone.

WOLFF, Kaspar Friedrich (1733–1794) : German anatomist. Professor of Anatomy and Physiology at St Petersburg. Helped establish modern embryology and the doctrine of the germ layers.
   Wolffian Duct (or Wolff's Duct) — the embryonic mesonephric duct running from the mesonephros to the cloaca.

WOOLNER, Thomas (1825–1892) : English sculptor and poet. Drew Charles Darwin's attention to the tubercle which now usually bears his name.
   Woolner's Tubercle — occasionally present on the helix of the pinna. (See Darwin.)

WORM, Ole (1588–1654) : Anatomist and theologian. Professor of Greek and Philosophy and later of Anatomy at Copenhagen.
   Wormian Bones — small bones found in the sutures of the skull, especially the lambdoid suture.

ZINN, Johann Gottfried (1727–1759) : Anatomist and botanist. Professor of Medicine and Director of the Botanical Gardens in Gottingen.
   Zonule of Zinn — the hyaloid membrane adjacent to the lens margin.

ZUCKERKANDL, Emil (1849–1910) : Professor of Anatomy at Graz and Vienna.
   Zuckerkandl's Fascia — the retro-renal fascia.